Brannigan's Building Construction for the Fire Service

Fourth Edition

Student Workbook

Francis L. Brannigan, SFPE
Founder of the New York City Fire Bell Club
Life Member of the National Fire Protection Association, International Association of Fire Chiefs, and International Society of Fire Service Instructors
Adjunct Instructor for the National Fire Academy, Emmitsburg, MD and the Maryland Fire and Rescue Institute, University of Maryland, College Park, MD

Glenn P. Corbett, PE
Associate Professor—Fire Science
John Jay College of Criminal Justice
Former Assistant Chief
Waldwick Fire Department, NJ

JONES AND BARTLETT PUBLISHERS
Sudbury, Massachusetts
BOSTON TORONTO LONDON SINGAPORE

Jones and Bartlett Publishers
World Headquarters
40 Tall Pine Drive
Sudbury, MA 01776
978-443-5000
info@jbpub.com
www.jbpub.com

Jones and Bartlett Publishers Canada
6339 Ormindale Way
Mississauga, Ontario L5V 1J2
Canada

Jones and Bartlett Publishers International
Barb House, Barb Mews
London W6 7PA
United Kingdom

National Fire Protection Association
1 Batterymarch Park
Quincy, MA 02169-7471
www.NFPA.org

Jones and Bartlett's books and products are available through most bookstores and online booksellers. To contact Jones and Bartlett Publishers directly, call 800-832-0034, fax 978-443-8000, or visit our website www.jbpub.com.

Substantial discounts on bulk quantities of Jones and Bartlett's publications are available to corporations, professional associations, and other qualified organizations. For details and specific discount information, contact the special sales department at Jones and Bartlett via the above contact information or send an email to specialsales@jbpub.com.

Editorial Credits/Author

Dan Smits, MBA, EFO
Crete Fire Department, IL

Production Credits

Chief Executive Officer: Clayton E. Jones
Chief Operating Officer: Donald W. Jones, Jr.
President, Higher Education and Professional Publishing: Robert W. Holland, Jr.
V.P., Sales and Marketing: William J. Kane
V.P., Production and Design: Anne Spencer
V.P., Manufacturing and Inventory Control: Therese Connell
Publisher, Public Safety Group: Kimberly Brophy
Senior Acquisitions Editor: William Larkin
Associate Managing Editor: Robyn Schafer
Production Editor: Karen Ferreira
Associate Photo Researcher and Photographer: Christine McKeen
Director of Marketing: Alisha Weisman
Cover Image: © The Sentinel, S.H. Wessel/AP Photos
Composition: Shepherd, Inc.
Text Printing and Binding: Courier Stoughton
Cover Printing: Courier Stoughton

ISBN-13: 978-0-7637-4879-1
ISBN-10: 0-7637-4879-X

6048

Printed in the United States of America
11 10 09 08 07   10 9 8 7 6 5 4 3 2 1

Table of Contents

Answer Key

Technology Resources

A key component to the teaching and learning system are the interactivities to help students learn building construction.

www.Fire.jbpub.com/Brannigan

Make full use of today's teaching and learning technology with www.Fire.jbpub.com/Brannigan. The site has been specifically designed to compliment *Brannigan's Building Construction for the Fire Service, Fourth Edition*. Some of the resources available include:

- Chapter Pretest tests your knowledge of the important concepts in each chapter and provides a page reference for each answer
- Vocabulary Explorer offering:
  - **Online Glossary** This online virtual dictionary lets you search or browse all of the key terms within the *Fourth Edition* in three ways: by term, alphabetically, or by chapter number.
  - **Animated Flashcards** Learning has never been so visually stimulating. Animated flashcards are a valuable resource to help you review your knowledge of key terms and concepts.
  - **Crossword Puzzles** A stimulating way to review key terms. From the clues provided, complete an online crossword puzzle using the key terms presented. If you get stuck, you can access different hints to help you on your way.

Workbook Activities

The following activities have been designed to help you. Your instructor may require you to complete some or all of these activities as a regular part of your training program. You are encouraged to complete any activity that your instructor does not assign as a way to enhance your learning.

Matching

Match each of the terms in the left column to the appropriate definition in the right column.

_______ **1.** Building fire

_______ **2.** Structural fire

_______ **3.** Prefire planning

_______ **4.** Mayday

_______ **5.** Construction type

_______ **6.** Interactive preplans

A. Reviewing a building ahead of a fire to gather information

B. Universal distress signal

C. Fire within a building's structure

D. Fire that includes a building's structure

E. Information that should be included in an officer's initial report upon arrival

F. Prospective form of real-time data, transmitted directly to responding firefighters

Chapter 1

Introduction

Multiple Choice

Read each item carefully, then select the best response.

_______ 1. Accurate language and correct terminology are important to fire fighters because:

A. they help avoid confusion.

B. they effectively convey needs in emergency situations.

C. they are signs of safety and professionalism.

D. all of the above.

_______ 2. Each year, approximately __________ fire fighters are killed in the line of duty.

A. 10

B. 100

C. 1000

D. None of the above

_______ 3. Prefire planning is a crucial component of firefighting because:

A. it provides specific and detailed data about immediate evacuation strategies.

B. it gives fire fighters the tools to prepare against potential hazards.

C. it relates the necessary jargon for common building constructions.

D. it outlines the rules about action and the chain of command.

NOTES

Short Answer

Complete this section with short written answers, using the space provided.

1. List the five main details that the first officer at a possible building fire should report.

___

___

___

NOTES

2. Identify three major components of prefire planning.

____

____

____

Workbook Activities

The following activities have been designed to help you. Your instructor may require you to complete some or all of these activities as a regular part of your training program. You are encouraged to complete any activity that your instructor does not assign as a way to enhance your learning.

Matching

Match each of the terms in the left column to the appropriate definition in the right column.

_______ **1.** Dead load

_______ **2.** Compression

_______ **3.** Tension

_______ **4.** Torsion

_______ **5.** Stress

_______ **6.** Strain

_______ **7.** Concentrated load

_______ **8.** Live load

_______ **9.** Eccentric load

_______ **10.** Lateral impact load

A. A twisting force

B. An internal force, measured in pounds per unit area, that resists a load

C. A direct pushing force, in line with the axis member

D. An internal force, percent of elongation that occurs when a material is stressed

E. A force that acts on a structure from a horizontal direction

F. The weight of a building

G. The weight of a building's contents

H. A force that is perpendicular to the plane of a section but does not pass through the center of the section

I. A pulling or stretching force, in line with the axis of the body

J. A load acting on a very small area of the structure's surface

Chapter 2

Concepts of Construction

Multiple Choice

Read each item carefully, then select the best response.

_______ 1. The wall that typically that has the highest fire rating and is the strongest within a building is a:

A. load-bearing wall.

B. fire wall.

C. curtain wall.

D. panel wall.

_______ 2. The type of connection that allows the weight of the building to hold them in place is a:

A. wet joint.

B. gusset connection.

C. gravity connection.

D. grillage.

_______ 3. There are numerous types of beams; the type that is supported at three or more points is a/an __________ beam.

A. continuous

B. cantilever

C. simple

D. overhanging

NOTES

_______ 4. The line along a beam that does not change is the __________ or plane.
A. vertical axis
B. horizontal axis
C. stiffness axis
D. neutral axis

_______ 5. The most effective shape for a column is one that:
A. distributes the material equally around the axis as far as possible from the center of the cylinder.
B. distributes the material equally around the axis as close as possible to the center of the cylinder.
C. distributes the material equally around the axis as far as possible from the edge of the cylinder.
D. distributes the material equally around the axis as close as possible to the edge of the cylinder.

_______ 6. A structural member that transmits a compressive force along a straight path in the direction of the member is called a:
A. column.
B. beam.
C. floor joist.
D. strut.

_______ 7. The __________ of a beam is the result of force exerted by a beam on a support.
A. load
B. reaction
C. release rate
D. compression

_______ 8. A __________ wall acts as one unit.
A. homogeneous
B. stable
C. voussoir
D. ledgered

_______ 9. Steel heated to 1000°F elongates __________ per 100 feet of length.
A. 7 inches
B. 8 inches
C. 9 inches
D. 10 inches

_______ 10. Which of the following are commonly used in heavy timber buildings?
A. Lightweight trusses
B. Fire-cut joists
C. Curtain walls
D. Self-releasing floors

NOTES

Labeling

Label the following diagram with the correct terms.

Load-Transferring Beams

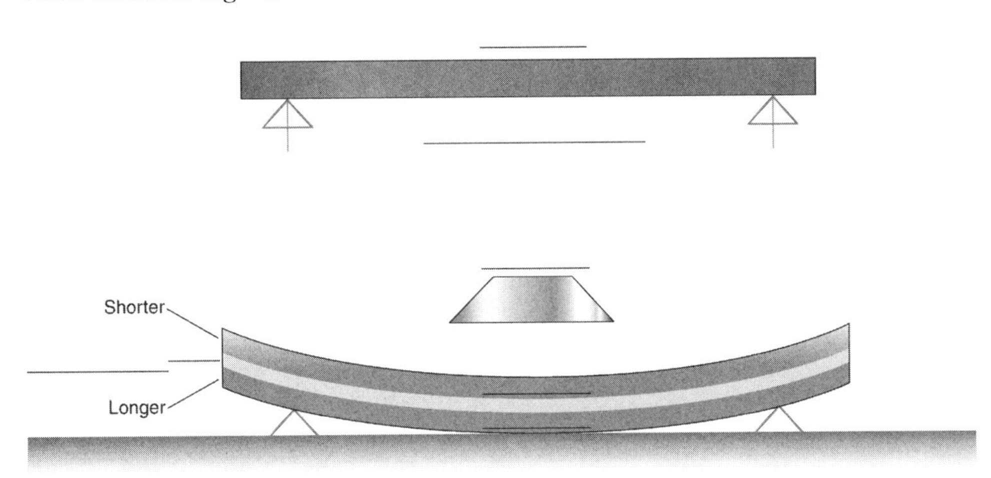

Fill-in-the-Blank

Read each item carefully, then complete the statement by filling in the missing word(s).

1. A/An __________ is made by sandwiching a piece of steel between two wooden beams.
2. The combination of two different materials in a floor is called a/an __________ floor.
3. Externally braced buildings are known as __________.
4. The rate of available energy released is the __________.
5. When changes are made in the foundation of an existing wall, it may be necessary to insert a/an __________.
6. A/An __________ is probably the oldest structural member.
7. Struts or rakers are other names for __________ columns.
8. A/An __________ beam moves loads laterally when it is not convenient to arrange columns in an ideal way.
9. It is important to know how loads are __________ from the point of application to the ground.
10. Some steel buildings have connections that redirect overloads to other sections of the building; this is known as a/an __________ design.

NOTES

True/False

Read each item carefully, then, if you believe the statement to be more true than false, write "T" in the space provided. If you believe the statement to be more false than true, write "F" in the space provided.

_____ 1. Vierendeel trusses are square trusses with weak corner bracings.

_____ 2. An axial load is a force that passes through the centroid of the section under consideration.

_____ 3. The best way to rate a building component is based on its ultimate strength.

_____ 4. Prefire plans should note the estimated fire load.

_____ 5. A typical sofa has a peak HRR of 1500 kW.

_____ 6. A steel beam resting on a masonry wall is an example of a concentrated load.

_____ 7. Suspended loads depend on interior columns for support.

_____ 8. The universal spacing for sawn wooden beams in ordinary construction is 12 inches.

_____ 9. A theater marquee is an example of a cantilever beam.

_____ 10. Steel beams are typically I-shaped.

Short Answer

Complete this section with short written answers, using the space provided.

1. What is the difference between panel (or curtain) walls and party walls?

____

____

____

2. What are the two general types of connections?

____

____

____

NOTES

3. How can a connection fail?

4. Name three methods of wall bracing.

5. What are Euler's Law columns?

CLUES

Across

3 A wedge-shaped block whose converging sides radiate from a center forming an element of an arch or vaulted ceiling

6 Bricks designed to join wythes

7 In place when a mass of masonry is placed against a wall to strengthen it

8 Braces a column diagonally

10 A masonry column, built on the inside surface of a wall

Down

1 The outside member of a truss, as opposed to inner webbed members

2 The end of a joist is cut off at an angle to permit the joist to fall out of the wall without damaging the wall

4 A truss that has very rigid corner bracing

5 A beam and column combined into a single element

9 Wall that divides tenant spaces

Word Fun

The following crossword puzzle is an activity provided to reinforce correct spelling and understanding of relevant terminology. Use the clues provided to complete the puzzle.

NOTES

Research Question

The following research question will give you an opportunity to explore the principles presented in *Brannigan's Building Construction for the Fire Service, Fourth Edition* in real situations within your community. Consider the following question and use the resources available to you to provide a detailed response on a separate piece of paper.

Identify five buildings of different occupancies within your community and, using the concepts of construction presented in this chapter, describe the various loads that may be factors in these buildings in the case of a fire.

Workbook Activities

The following activities have been designed to help you. Your instructor may require you to complete some or all of these activities as a regular part of your training program. You are encouraged to complete any activity that your instructor does not assign as a way to enhance your learning.

Matching

Match each of the terms in the left column to the appropriate definition in the right column.

_______ **1.** Area of refuge
_______ **2.** Malleable
_______ **3.** Pile
_______ **4.** Folded Plate
_______ **5.** Falsework
_______ **6.** Corrugation
_______ **7.** Fire partition
_______ **8.** Thermosets
_______ **9.** Thermoplastics
_______ **10.** Admixtures

A. Grooved, ridged material
B. Wall assembly that subdivides a building to prevent the spread of fire
C. Type of material that can drip to produce secondary fires
D. Space separated by smoke-rated barrier in which a tenable environment is maintained during a fire
E. Placed in concrete to give it special characteristics
F. Used to strengthen a roof over large areas
G. Temporary shoring or lateral bracing to support the work in process of construction
H. Material that chars and burns but does not flow
I. Property of a metal that allows it to be shaped
J. Large series of large timbers or steel driven into the ground

Chapter 3

Methods and Materials of Construction, Renovation, and Demolition

Multiple Choice

Read each item carefully, then select the best response.

_______ 1. Which type of wood panel has all strands laid at right angles?
- **A.** Oriented strand board
- **B.** Plywood
- **C.** Timber board
- **D.** Crazed strand board

_______ 2. Which stone is particularly subject to spalling when exposed to fire?
- **A.** Marble
- **B.** Limestone
- **C.** Granite
- **D.** Sandstone

_______ 3. At what temperature does structural steel fail?
- **A.** 500°F
- **B.** 750°F
- **C.** 1000°F
- **D.** 1500°F

NOTES

_______ **4.** If a material contributes fuel to the combustion process, it is considered:

A. fire-resistant.
B. combustible.
C. admixture.
D. aggregate.

_______ **5.** Which material is strong in tension and weak in compression?

A. Concrete
B. Wood
C. Hemp rope
D. Steel

_______ **6.** Which of the following is used to support smaller buildings and those built on stronger soils?

A. Footings
B. Piles
C. Caissons
D. Stilts

_______ **7.** Soil walls are protected against collapse by all of the following, EXCEPT:

A. crosslot bracing.
B. rakers.
C. tiebacks.
D. tie braker.

_______ **8.** All of the following are construction regulations, EXCEPT:

A. plumbing code.
B. electrical code.
C. heating and air conditioning code.
D. mechanical code.

_______ **9.** Which of the following people is typically responsible for the overall project?

A. Civil engineer
B. General contractor
C. Architect
D. Structural engineer

_______ **10.** Which of the following presents a common safety danger at a construction site?

A. Unsecured openings
B. Missing stairwells
C. Moving objects
D. All of the above

NOTES

Fill-in-the-Blank

Read each item carefully, then complete the statement by filling in the missing word(s).

1. A/An __________ code regulates the actual design and construction of new buildings.
2. A/An __________ code regulates the activities that take place in existing buildings.
3. __________ laws dictate land usage in specific cities.
4. The __________ is a set of regulations passed in 1990 that includes, among other things, regulations requiring areas of refuge.
5. __________ is the organization that enforces regulations related to safety on construction sites.
6. The __________ is responsible for the design of a building and decides how a building will perform.
7. A fire protection __________ designs the fire protection systems of a building.
8. The __________engineer designs the layout of the site, including parking, drainage, and roadways.
9. A/An __________ engineer designs heating, air conditioning, and plumbing systems.
10. The __________engineer designs the lighting, telecommunications, and emergency power systems.

True/False

Read each item carefully, then, if you believe the statement to be more true than false, write "T" in the space provided. If you believe the statement to be more false than true, write "F" in the space provided.

_____ 1. Buildings under construction are usually more dangerous than completed structures.

_____ 2. A fire code provides the maximum levels of health and safety.

_____ 3. While on site, firefighters should test private hydrants.

_____ 4. Progressive collapses can be stopped by shear walls.

_____ 5. Dimensional lumber is between 5 and 7 inches in nominal thickness.

_____ 6. Cast iron is a very strong material.

_____ 7. A slurry wall is made up of engineered wood products.

_____ 8. Columns typically form a very strong support.

NOTES

_____ 9. Fire retardants do not render wood noncombustible.

_____ 10. Wet wood has less strength than dry wood.

Short Answer

Complete this section with short written answers, using the space provided.

1. What hazards may be present in a standpipe system in a high-rise building construction project?

____

____

____

2. What laws dictate land usage?

____

____

____

3. What act mandates the construction of areas of refuge in multistory buildings?

____

____

____

4. What professionals may be involved in the building design and construction process?

____

____

____

5. What are the negative characteristics of steel?

____

____

____

NOTES

Fire Alarms

The following case scenarios will give you an opportunity to explore the concerns associated with building construction for the fire service. Read each scenario, then answer each question in detail.

1. In many buildings, glass has been fabricated to withstand severe circumstances such as blasts or hurricanes. What issues do these special types of glass prevent for firefighters?

____

____

____

____

____

2. A building that is being constructed of concrete has several concerns for firefighters. What are they?

____

____

____

____

____

Research Question

The following research question will give you an opportunity to explore the principles presented in *Brannigan's Building Construction for the Fire Service, Fourth Edition* in real situations within your community. Consider the following question and use the resources available to you to provide a detailed response on a separate piece of paper.

Identify a building under renovation within your community. Describe the methods and materials of construction present in the building and any hazards that may be unique to the building due to the renovation.

Workbook Activities

The following activities have been designed to help you. Your instructor may require you to complete some or all of these activities as a regular part of your training program. You are encouraged to complete any activity that your instructor does not assign as a way to enhance your learning.

Matching

Match each of the terms in the left column to the appropriate definition in the right column.

_______ **1.** Exit access

_______ **2.** Exit discharge

_______ **3.** Exit

_______ **4.** Phase I operation

_______ **5.** Phase II operation

_______ **6.** Class A Flame Spread Rating

_______ **7.** Class B Flame Spread Rating

_______ **8.** Class C Flame Spread Rating

_______ **9.** Building code

_______ **10.** Fire code

A. Operation in which elevators are recalled (usually) to the first floor

B. Rating index of 76–200

C. Operation in which a firefighter selects one of the elevators to access upper floors

D. Actual path from any point in a building to a fire-rated stairwell

E. Rating index of 0–25

F. Rated stairwell or exit passageway

G. Rating index of 26–75

H. Door to the exterior

I. Regulates the activities that take place in buildings

J. Regulates the design and construction of new buildings

Chapter 4

Building and Fire Codes

Multiple Choice

Read each item carefully, then select the best response.

_______ 1. The occupant load factor used for swimming pool decks is:
- A. 15 square feet per person.
- B. 30 square feet per person.
- C. 35 square feet per person.
- D. 50 square feet per person.

_______ 2. The Americans with Disabilities Act of 1990 outlined several provisions increasing the safety of the disabled, including all of the following, EXCEPT:
- A. insulated stairwell landings.
- B. fire alarm systems with both visual and audible signaling devices.
- C. wheelchair ramps.
- D. areas of refuge.

_______ 3. According to Table 4–2, NFPA 5000 allows a maximum height of _______ for Type II construction.
- A. 65 feet
- B. 85 feet
- C. 160 feet
- D. 180 feet

NOTES

_______ **4.** According to Table 4–2, the maximum number of stories for a Residential Type I building is __________ stories.

A. 100

B. 150

C. 350

D. unlimited

_______ **5.** According to Table 4–2, what is the fire resistance rating requirement for exterior walls in a Type II building?

A. 1

B. 2

C. 3

D. 4

_______ **6.** Building codes are more restrictive on more dangerous occupancy types, which include:

A. jails.

B. hospitals.

C. homes for the elderly.

D. all of the above.

_______ **7.** According to Table 4–1, floor-ceiling assemblies in a Type I building should be rated for __________ hour(s).

A. 1

B. 2

C. 3

D. 4

_______ **8.** The standard for the installation of standpipes is:

A. NFPA 13.

B. NFPA 14.

C. NFPA 72.

D. NFPA 1710.

_______ **9.** According to the *International Building Code*, Type IV is what type of construction?

A. Fire resistive

B. Ordinary

C. Heavy timber

D. Wood frame

_______ **10.** The minimum size of an emergency escape window is:

A. 5.0 square feet.

B. 5.7 square feet.

C. 6.0 square feet.

D. 6.3 square feet.

NOTES

Fill-in-the-Blank

Read each item carefully, then complete the statement by filling in the missing word(s).

1. Occupant load factors are expressed in __________ area, unless marked "net."
2. The model codes allow sprinkler __________, which lessen code requirements for some buildings.
3. NFPA 5000 classifies a restaurant with 1000 occupants as a/an __________.
4. Under the *International Building Code*, a church would be a/an __________ occupancy.
5. The older model codes are called the __________ codes.
6. The *Standard Test Method for Surface Burning Characteristics of Building Materials* is abbreviated __________.
7. The three-digit number used in the NFPA construction types stands for the __________ of each type of construction.
8. __________ buildings are ones that do not easily fit into an existing classification.
9. The use of fixed burglar bars is __________ in all model codes.
10. The Steiner Tunnel Test results in a flame spread __________.

True/False

Read each item carefully, then, if you believe the statement to be more true than false, write "T" in the space provided. If you believe the statement to be more false than true, write "F" in the space provided.

_____ 1. Standpipe hose valves are often tied to the locations of exits.

_____ 2. There are four components to the means of egress.

_____ 3. Panic hardware is required in all businesses and storage facilities.

_____ 4. The standard for elevators is ASME A17.1.

_____ 5. There are five types of construction.

_____ 6. Building codes regulate the area in square feet per floor before fire walls are required.

_____ 7. Model codes are regulatory documents.

_____ 8. One of the most important test standards is ASTM E-117.

_____ 9. The International Code Council and NFPA prepare both building and fire codes.

_____ 10. It has been said that 90 percent of a building code deals with fire safety.

NOTES

Short Answer

Complete this section with short written answers, using the space provided.

1. What types of construction are regulated in the International Building Code?

2. What are the two varieties of elevator lifting mechanisms?

3. What requirements do building codes set for emergency egress?

4. What factors are considered in setting the requirements for emergency egress?

5. What are elevator door restrictors?

NOTES

Fire Alarms

The following case scenarios will give you an opportunity to explore the concerns associated with building construction for the fire service. Read each scenario, then answer each question in detail.

1. A local architect has asked you for the requirements in a Type II subcategory 222 building for the exterior bearing walls. What are the hourly requirements? (Use the table on page 68 to help you.)

____

____

____

____

____

2. Within a sprinklered Type III building, the area per floor and number of stories allowed for a storage occupancy of ordinary hazard is regulated by the allowable height and area tables. What is the allowable building area per floor and number of stories? (Use the table on page 71 to help you.)

____

____

____

____

____

Research Question

The following research question will give you an opportunity to explore the principles presented in *Brannigan's Building Construction for the Fire Service, Fourth Edition* in real situations within your community. Consider the following question and use the resources available to you to provide a detailed response on a separate piece of paper.

Identify the adopted codes in the place where you live. Include the name of the codes and the year they were adopted. Try to obtain a copy of the adoption ordinance or state statute for the building or fire code. Remember that there are many different types of codes, including those for plumbing, mechanical, fire, electrical, and property maintenance. Identify all of the codes that are applicable to new construction.

Workbook Activities

The following activities have been designed to help you. Your instructor may require you to complete some or all of these activities as a regular part of your training program. You are encouraged to complete any activity that your instructor does not assign as a way to enhance your learning.

Matching

Match each of the terms in the left column to the appropriate definition in the right column.

_______ **1.** Remote annunciators
_______ **2.** Matchboarding
_______ **3.** Gravity vent
_______ **4.** Damper
_______ **5.** Demand area
_______ **6.** Zone
_______ **7.** Steiner Tunnel Test
_______ **8.** Marble
_______ **9.** Flameover
_______ **10.** Combustible acoustical tile

A. Fiberboard punched with holes
B. Small panels indicating the fire detector activation
C. Gallons per minute per square foot required within a sprinkler system
D. Location in buildings where dangerous gases are handled
E. Another name for *NFPA 255: Standard Method of Test Surface Burning Characteristics of Building Materials*
F. Rapid spread of flame over one or more surfaces
G. Individual riser segments, one on top of the other
H. Valve or plate for controlling the draft or flow of gases
I. Material often used in older library flooring
J. Ceilings made of embossed steel and wooden boards

Chapter 5

Features of Fire Protection

Multiple Choice

Read each item carefully, then select the best response.

_______ 1. Which building type is described in building codes but is not technically accurate?
- A. Combustible
- B. Frame
- C. Noncombustible
- D. Heavy timber

_______ 2. Which class of standpipe system has a 2.5-inch (65-mm) connection?
- A. Class I
- B. Class II
- C. Class III
- D. Class IV

_______ 3. A type of sprinkler that dumps most of the water on the fire is called a/an:
- A. ceiling upright.
- B. early suppression fast response (ESFR).
- C. sidewall.
- D. water flow switch.

NOTES

_______ **4.** A device that is usually located near an entrance to a building to monitor the building alarm system is called a/an:

A. entrance detection station.
B. remote FACP.
C. initiating device.
D. remote annunciator.

_______ **5.** All of the following components are preplanning sprinkler considerations, EXCEPT:

A. total water supply demand of the system.
B. location of manual initiation for postaction and deluge systems.
C. fire department connection location and the areas it serves.
D. types of heads, including any unusual characteristics.

_______ **6.** All of the following are standpipe water supply classifications, EXCEPT:

A. automatic-wet.
B. semiautomatic-dry.
C. semiautomatic-wet.
D. manual-wet.

_______ **7.** Site development incentives for sprinklers may include all of the following, EXCEPT:

A. street width reduction.
B. fewer hydrants.
C. smaller supply pipe.
D. fewer subdivision entrances.

_______ **8.** The most expensive byproduct of fire suppression is __________ damage.

A. water
B. fire
C. utility
D. smoke

_______ **9.** Which gas is now considered as dangerous as carbon monoxide in fires?

A. Hydrogen dioxide
B. Cyanide hydroxide
C. Hydrogen cyanide
D. Phenol

_______ **10.** What does the radiant flux test measure about a material?

A. Its ability to resist flame spread
B. The time it takes to ignite after contact with flame
C. The degree to which it is usually damaged by firefighting stream
D. The amount of smoke it emits when enflamed

NOTES

Fill-in-the-Blank

Read each item carefully, then complete the statement by filling in the missing word(s).

1. Fire resistance that is based on the testing of a wall floor or column assembly is called __________ fire resistance.

2. Fire-retardant surface coating is effective only if applied as __________.

3. __________ rule states that any exposure in which the concentration multiplied by minutes exposed equals 33,000 is likely to be dangerous.

4. There are two types of closure devices for doors: automatic and __________.

5. Sprinkler valves are made to indicate their position; if the stem is protruding, it is __________.

6. __________ cans may act as flying rockets in a fire, so spaces where they are stored require a specialized sprinkler design.

7. Smoke control systems will have some type of __________ panel.

8. Automatic closing doors operate by melting of the __________ link.

9. Fire fighters should make sure fire doors are __________ before advancing through them.

10. Horizontal exits usually have a/an __________-hour rated wall.

True/False

Read each item carefully, then, if you believe the statement to be more true than false, write "T" in the space provided. If you believe the statement to be more false than true, write "F" in the space provided.

_____ 1. One of the most important elements of life safety is proper egress.

_____ 2. The term "inflammable" is gaining popularity since the NFPA approved its use.

_____ 3. Inherent fire resistance is a structural member's resistance to collapse by fire because of the nature of its material or assembly.

_____ 4. It is not possible to train in the same way for collapse and hidden hazards.

_____ 5. Fire growth can be differentiated by the location characteristics of being hidden or exposed.

_____ 6. Flame spread is greatly influenced by building construction.

_____ 7. Plaster is extremely flammable and contributes to a fire.

_____ 8. Estimation should be made of the potential for flame spread over interior finishes.

NOTES

_____ 9. Asphalt-covered steel is called Oregonian protected metal.

_____ 10. Typically, it takes about 15 minutes to reach flashover in most fires.

Short Answer

Complete this section with short written answers, using the space provided.

1. How does interior finish increase a fire hazard?

____

____

____

2. What building code areas aim to reduce the chance of conflagrations?

____

____

____

3. What building code areas aim to limit spread of fire within a building?

____

____

____

4. What indications of imminent collapse have been shown to be inadequate?

____

____

____

5. What is the purpose of sprinkler systems equipped with cycling sprinklers?

____

____

____

Word Fun

The following crossword puzzle is an activity provided to reinforce correct spelling and understanding of relevant terminology. Use the clues provided to complete the puzzle.

CLUES

Across

3 Fiberglass or rock-wool insulation with various thicknesses available

4 Venting of smoke

7 Type of gaseous fire-extinguishing-agent system that does not leave a residue

8 System used to protect materials damaged by water

9 Subdivision of a building into small areas so that fire or smoke is confined

Down

1 Fiberboard in which holes have been punched

2 Burning of heated, gaseous products of combustion after oxygen is introduced into a space where oxygen had been depleted

3 Low-density fiberboard made of wood fibers or sugar cane residue

5 Type of switch that, when flipped, silences the fire alarm but leaves it activated

6 Resistant to fire

NOTES

Fire Alarms

The following case scenarios will give you an opportunity to explore the concerns associated with building construction for the fire service. Read each scenario, then answer each question in detail.

1. A fire is located in the first floor of a department store that has three levels. The building has elevators, escalators, and sprinklers. What may be a concern about the sprinklers and fire protection systems in this building?

____

____

____

____

____

2. When pre-planning a smoke management system, there are several things to keep in mind. Identify the issues of pre-planning for these systems and how they might affect your firefighting considerations.

____

____

____

____

____

NOTES

Research Question

The following research question will give you an opportunity to explore the principles presented in *Brannigan's Building Construction for the Fire Service, Fourth Edition* in real situations within your community. Consider the following question and use the resources available to you to provide a detailed response on a separate piece of paper.

Fire protection involves both active and passive types of systems. Pick a structure within your community that has what you would consider to be some of the more elaborate fire protection systems in it. Describe the systems in place, including the advantages and disadvantages to a firefighting situation that may occur. Describe any systems that may be impaired without your knowledge.

Workbook Activities

The following activities have been designed to help you. Your instructor may require you to complete some or all of these activities as a regular part of your training program. You are encouraged to complete any activity that your instructor does not assign as a way to enhance your learning.

Matching

Match each of the terms in the left column to the appropriate definition in the right column.

_______ **1.** Header
_______ **2.** Parallel chord
_______ **3.** Rafter
_______ **4.** Soffit
_______ **5.** Splines
_______ **6.** Balloon frame
_______ **7.** Chamfer
_______ **8.** Engineered wood
_______ **9.** Matched lumber
_______ **10.** Glued laminated timbers

A. Wooden strips that fit into grooves in two adjacent planks to make a tight floor
B. Joist that parallels floor/roof beams to create an opening
C. False spaces above built-in cabinets
D. To cut off the corners of a timber to retard ignition
E. A truss where the chords are parallel
F. Wooden structure in which all vertical studs in the exterior walls extend the full height of the frame
G. Planks fastened together to form a solid timber
H. Members used to support the roof sheeting and loads
I. Wood modified from its original state
J. Tongue and grooved lumber

Chapter 6

Wood Frame Construction

Multiple Choice

Read each item carefully, then select the best response.

_______ **1.** In which type of truss does the compression member extend downward?

A. Queen post truss
B. Parallel chord truss
C. Triangular truss
D. Inverted king post truss

_______ **2.** Which structural member is placed first on a foundation?

A. Base plate
B. Sill
C. Primary member
D. Joist

_______ **3.** A lightweight truss substitutes:

A. mass for geometry.
B. height for length.
C. geometry for mass.
D. length for height.

NOTES

_______ **4.** Rising roofs in truss construction can cause all of the following, EXCEPT:

A. damaged gas lines.
B. weakened joists.
C. pyrolysis.
D. loose gusset plates.

_______ **5.** Which of the following is a component of truss-framed construction?

A. A unitized frame
B. A large void space
C. Multiple trusses forming a hip
D. Slow construction speed

_______ **6.** Treated wood may be used for all of the following reasons, EXCEPT:

A. to resist insect damage.
B. to create a fire-retardant product.
C. to avoid water damage.
D. to avoid toxicity when burned.

_______ **7.** Which type of girder is made up of a steel plate or plywood sandwiched between two beams?

A. Bearing plate
B. Flitch plate
C. Dual plate
D. Sandwich plate

_______ **8.** Which of the following is a problem with corrugated metal siding?

A. Quick collapse potential
B. Difficult forced entry issues
C. Severe electrical hazards
D. Severe structural weaknesses

_______ **9.** What is the biggest hazard of low-density fiberboard?

A. Vermin
B. Moisture
C. Fire
D. Delamination

_______ **10.** Terms associated with mortise and tenon joints include all of the following, EXCEPT:

A. socket.
B. tongue.
C. trunnel.
D. bridging.

NOTES

Fill-in-the-Blank

Read each item carefully, then complete the statement by filling in the missing word(s).

1. In balloon framing, studs run two or more stories high, from the __________ to the eave line.
2. A framed building is understood to have a skeleton of beams and columns in which the walls are just __________ walls.
3. __________ is a serious threat in a balloon-framed building.
4. Trusses can provide huge clear spans at a/an __________ weight considerably less than that of a corresponding beam.
5. Tactics based on __________ joist floors will kill fire fighters if used on buildings with truss floors.
6. Draftstops typically limit __________ spread of fire.
7. Firestops typically limit __________ spread of fire.
8. Plank-like sections of nominal two-inch or thinner boards glued under pressure to form arches are known as __________.
9. __________ joints are made by cutting a series of long points into the end of two-by-fours before gluing them together.
10. Plywood siding is typically delivered in __________-foot wide sheets.

True/False

Read each item carefully, then, if you believe the statement to be more true than false, write "T" in the space provided. If you believe the statement to be more false than true, write "F" in the space provided.

_____ 1. Tongue-and-groove roof planks are a resourceful method to cover rafters.

_____ 2. Brick structures have so-called "good press."

_____ 3. Laminated wood beams can delaminate easily.

_____ 4. NFPA 13R sprinkler systems are partial sprinkler systems.

_____ 5. The failure of one element of a truss may cause the entire truss to fail.

_____ 6. The top chord of a truss is under tension.

_____ 7. Modern high-rise buildings weigh as little as 20 pounds per square foot.

_____ 8. The compressive connecting members of a truss are called struts.

NOTES

_____ **9.** Bridging is laid on top of floor joists.

_____ **10.** Many post-and-frame buildings are enclosed with foam core panels.

Short Answer

Complete this section with short written answers, using the space provided.

1. What are the two different types of post trusses?

____

____

____

2. What are the three parts of the web in a truss?

____

____

____

3. What are the two types of firestopping?

____

____

____

4. What is the problem associated with plywood?

____

____

____

5. Describe the ratings of roofing materials.

____

____

____

Word Fun

The following crossword puzzle is an activity provided to reinforce correct spelling and understanding of relevant terminology. Use the clues provided to complete the puzzle.

CLUES

Across

4 A mineral known as __________ is commonly used as insulation.

5 A vertical compression member that supports loads is called a/an __________.

7 A/An __________ board is placed on the top of a roof onto which the upper ends of the rafters are fastened.

8 A/An __________ is used in truss systems and is subject to tensile and/or pulling forces.

Down

1 Truss systems are connected together by the use of a/an __________.

2 The points at which connections occur in a truss system are called __________.

3 A/An __________ is the rafter at an angle where two sloping roofs or sides of a roof meet.

4 The lower slope of a roof that is formed by the connection of two inclined planes is called a/an __________.

6 In order to support the header in a floor opening, a/an __________ is installed.

8 Wooden pegs, also known as a/an __________, are used to pin together mortise and tenon joints.

NOTES

Fire Alarms

The following case scenarios will give you an opportunity to explore the concerns associated with building construction for the fire service. Read each scenario, then answer each question in detail.

1. Construction has just begun on a local plot where a major fast-food chain is building a new, wooden building. What dangers may this present?

____

____

____

____

____

2. A house built in 1938 is on fire. What type of construction is this building most likely to be, and what hazards could this construction present to fire fighters?

____

____

____

____

____

NOTES

Research Question

The following research question will give you an opportunity to explore the principles presented in *Brannigan's Building Construction for the Fire Service, Fourth Edition* in real situations within your community. Consider the following question and use the resources available to you to provide a detailed response on a separate piece of paper.

Identify a lumberyard within the area of your community. What types of wood products does it carry? Are there any unusual or trendy wood products entering the local construction market? Identify some of the hazards and try to assemble examples of some of the products for training sessions at your fire station.

Workbook Activities

The following activities have been designed to help you. Your instructor may require you to complete some or all of these activities as a regular part of your training program. You are encouraged to complete any activity that your instructor does not assign as a way to enhance your learning.

Matching

Match each of the terms in the left column to the appropriate definition in the right column.

_______ **1.** Cast iron box

_______ **2.** Conflagration breeder

_______ **3.** Corbelled

_______ **4.** Fire cut

_______ **5.** Type IV construction

_______ **6.** Mill

_______ **7.** Scupper

_______ **8.** Vacant buildings

_______ **9.** Slow burning

_______ **10.** Sprinkler system

A. The end of a joist cut at an angle to permit the joist to fall out of a wall without acting like a lever

B. A series of projections in a wall

C. Only defense against a raging fire in an old mill of heavy timber construction

D. The earliest form of heavy timber construction

E. An outlet for drainage in a wall for a roof

F. Classification for heavy timber construction

G. A multi-story building with exposure problems that can create an uncontrollable fire

H. Built into a wall to receive a girder

I. A significant fire hazard

J. Characteristic of a building that should allow a fire in that building to be brought under control before the building itself becomes involved

Chapter 7

Heavy Timber and Mill Construction

Multiple Choice

Read each item carefully, then select the best response.

_______ 1. Mill features include all of the following, EXCEPT:

A. interior bearing and nonbearing walls with solid brick or stone masonry.

B. columns and beams made of heavy timber with cast iron connectors.

C. thick grooved, splined, or laminated floor planks.

D. roof planks supported by beams or timber arches and trusses.

_______ 2. Mill features include all of the following, EXCEPT:

A. openings between floors enclosed by adequate fire barriers.

B. ends of girders that are fire cut.

C. cast iron box built into the attic to receive the end of a girder.

D. scuppers or drains in the wall to drain off water.

_______ 3. Mill features include all of the following, EXCEPT:

A. waterproof floors.

B. division into sections by firewalls.

C. absence of concealed spaces.

D. absence of an automatic sprinkler system.

NOTES

_______ **4.** Typical departures from the true mill construction concept include all of the following, EXCEPT:

A. cast iron or unprotected steel columns.
B. dropped ceilings.
C. steel or part-steel trusses.
D. unprotected vertical openings.

_______ **5.** A fully involved, multi-story heavy timber building is a conflagration breeder for all of the following reasons, EXCEPT:

A. tremendous amount of radiant heat from each flaming window opening.
B. production of numerous large firebrands.
C. confined collapse zones.
D. shut-off sprinkler and standpipe systems.

True/False

Read each item carefully, then, if you believe the statement to be more true than false, write "T" in the space provided. If you believe the statement to be more false than true, write "F" in the space provided.

_____ **1.** The protection of vertical openings is vital in heavy timber construction.

_____ **2.** Concealed spaces are prevalent in heavy timber construction.

_____ **3.** The minimum dimension typically required by the code is 5 inches by 1 inch.

_____ **4.** The most common use for heavy timber construction today is in churches.

_____ **5.** Heavy timbers are ignited easily.

Short Answer

Complete this section with short written answers, using the space provided.

1. When is slow burning an advantage?

____

____

____

2. Why do vacant buildings pose a significant fire hazard?

____

____

____

NOTES

Research Question

The following research question will give you an opportunity to explore the principles presented in *Brannigan's Building Construction for the Fire Service, Fourth Edition* in real situations within your community. Consider the following question and use the resources available to you to provide a detailed response on a separate piece of paper.

Identify a heavy timber construction building in your area that may have had multiple uses. Do some research on the uses of the building since the building was built and describe how these changes may have affected the building conditions.

Workbook Activities

The following activities have been designed to help you. Your instructor may require you to complete some or all of these activities as a regular part of your training program. You are encouraged to complete any activity that your instructor does not assign as a way to enhance your learning.

Matching

Match each of the terms in the left column to the appropriate definition in the right column.

_______ **1.** Adobe

_______ **2.** Ashlar masonry

_______ **3.** Cockloft

_______ **4.** Course

_______ **5.** Dog iron

_______ **6.** Girder

_______ **7.** Bond course

_______ **8.** Mezzanine

_______ **9.** Pintle

_______ **10.** Parging

A. Void space between the top floor ceiling and the roof

B. Connects the girders and imparts some lateral stability under normal conditions

C. Stone cut in rectangular units

D. Bricks laid so the end is visible

E. Large, roughly molded, sun-dried clay units of varying sizes

F. A low-ceiling story located between two main stories

G. Square metal device used to transfer loads of columns on upper floors by passing the loads through intervening beams and girders

H. A horizontal line of masonry

I. Application of mortar to the back of the facing material

J. Large or principal beam of wood or steel used to support concentrated loads

Chapter 8

Ordinary Construction

Multiple Choice

Read each item carefully, then select the best response.

_______ 1. The chief common characteristic of ordinary construction is:
- **A.** the exterior walls are made of brick veneer.
- **B.** the exterior walls are masonry load-bearing walls.
- **C.** the floors are made of concrete.
- **D.** the buildings do not contain brick or wood joists.

_______ 2. Which wall separates two buildings and is meant to stop a fire?
- **A.** Fire partition
- **B.** Fire wall
- **C.** Fire barrier
- **D.** Bearing wall

_______ 3. What type of connections must be made whenever an opening is made in a wooden floor?
- **A.** Wall to beam
- **B.** Floor to beam
- **C.** Beam to beam
- **D.** Beam to column

_______ 4. Where a column is offset, the girder on which it rests is called a/an:
- **A.** transfer beam.
- **B.** pintle.
- **C.** offset beam.
- **D.** axial beam.

NOTES

_______ **5.** In order to get floor beams level in a masonry building, a wooden beam is laid in the brick wall at the line of the bottom of the floor beams; this can produce a:

A. void space of varying degrees in the wall.

B. stronger wall.

C. weak connection point.

D. plane of weakness.

_______ **6.** When heated to 1000°F, steel __________, which can lead to collapse.

A. hardens

B. shrinks

C. elongates

D. stiffens

_______ **7.** In what state is additional wall bracing very common as a part of the initial installation in a new building?

A. Minnesota

B. New York

C. Illinois

D. California

_______ **8.** Masonry walls in ordinary construction may consist of all of the following, EXCEPT:

A. stone.

B. terra cotta tile.

C. steel.

D. adobe.

_______ **9.** Which of the following is an example of ordinary construction?

A. Maximum-security prisons

B. High-rise apartment complex

C. One-story strip mall

D. Row of townhouses

_______ **10.** Problems of ordinary construction include which of the following?

A. Structural stability of the masonry wall

B. Stability of the interior column, girder, and beam system

C. Void spaces

D. All of the above

Fill-in-the-Blank

Read each item carefully, then complete the statement by filling in the missing word(s).

1. If __________ is falling off, it may indicate brick problems.

2. If any arch has a vousoir out, there is no __________.

NOTES

3. All __________ are inherently unstable.

4. __________ gas trapped in unvented voids can detonate violently and blow down walls.

5. Masonry walls are not designed to resist __________ impact loads.

6. The thrust of an arch is __________.

7. A/An __________ is a structure of wood, metal, or masonry that tops a wall and projects from it.

8. __________ walls are structural walls that are common to two buildings.

9. Fire doors should be noted on __________ plans.

10. Fire units should never pass through an overhead rolling fire door without __________ it.

True/False

Read each item carefully, then, if you believe the statement to be more true than false, write "T" in the space provided. If you believe the statement to be more false than true, write "F" in the space provided.

_____ 1. Void spaces are an inherent part of ordinary construction.

_____ 2. A simple wood beam floor is practical up to a limit of about 25 feet in width.

_____ 3. Ordinary construction is Type II construction.

_____ 4. There is usually good fire separation in ordinary constructed buildings.

_____ 5. The history of a building is very important.

_____ 6. Hollow and cavity walls limit penetration by rain.

_____ 7. Hollow walls of terra cotta tile present special hazards.

_____ 8. Parging may be a good indicator of poor brick.

_____ 9. Sand-lime mortar provides good adhesion with brick and withstands weather well.

_____ 10. A horizontal crack in a wall may indicate the wall is being pushed out by beams.

NOTES

Short Answer

Complete this section with short written answers, using the space provided.

1. What are some factors of ordinary construction that can lead to collapse?

____

____

____

2. What are the two ways to carry the walls above an opening?

____

____

____

3. What types of materials are used to make columns?

____

____

____

4. What items indicate a braced wall?

____

____

____

5. What is an example of an eccentric load on an ordinarily constructed building?

____

____

____

Word Fun

The following crossword puzzle is an activity provided to reinforce correct spelling and understanding of relevant terminology. Use the clues provided to complete the puzzle.

CLUES

Across

5 Arch in which a steel tension rod ties the ends of the arch together

6 Wall with a masonry facing that is not bonded but is attached to a wall so as to form an integral part of the wall

7 Rough stones of irregular shapes and sizes used in un-coursed masonry work

8 Beam used to laterally relocate the vertical load of columns to clear an open area

10 Masonry pier at a distance from the wall and connected to it that resists the outward thrust of the wall

Down

1 Any wall set at a right angle to any other wall

2 Single, continuous, vertical wall of bricks

3 Small court placed in large buildings

4 Bricks laid so the long side is visible

9 Older code provision that would not allow a structure to be built without the use of exterior masonry walls

NOTES

Fire Alarms

The following case scenarios will give you an opportunity to explore the concerns associated with building construction for the fire service. Read each scenario, then answer each question in detail.

1. You have been asked to survey buildings in your area. What are the fireground safety issues you definitely want to document and communicate to your department when discussing ordinary construction buildings?

____

____

____

____

____

2. You have been assigned to watch for signs of collapse during a structure fire. What are some general indicators of collapse?

____

____

____

____

____

Research Question

NOTES

The following research question will give you an opportunity to explore the principles presented in *Brannigan's Building Construction for the Fire Service, Fourth Edition* in real situations within your community. Consider the following question and use the resources available to you to provide a detailed response on a separate piece of paper.

Suspended loads are common on the outside of ordinary construction. Identify some buildings in your area with suspended loads that may be a concern for fire fighters.

Workbook Activities

The following activities have been designed to help you. Your instructor may require you to complete some or all of these activities as a regular part of your training program. You are encouraged to complete any activity that your instructor does not assign as a way to enhance your learning.

Matching

Match each of the terms in the left column to the appropriate definition in the right column.

________	**1.** Bar joist	**A.**	Beams set at right angles to trusses or roof rafters to provide support
________	**2.** Bulb	**B.**	Embedded into the surface
________	**3.** Bulkhead	**C.**	Void space made by utilizing deep parallel-chord trusses
________	**4.** Interstitial space	**D.**	Members with a Z-shaped cross section
________	**5.** Peened	**E.**	A tee where the end of the cutoff is thickened
________	**6.** Purlins	**F.**	Upright partition that divides a ship into compartments
________	**7.** Rakers	**G.**	A horizontal beam that ties rows of soldier beams together
________	**8.** Tube	**H.**	Diagonal columns that brace an entire structure
________	**9.** Waler	**I.**	Steel structural member rolled in cylindrical shapes
________	**10.** Zee	**J.**	Forms a long-span system used as floor supports

Chapter 9

Noncombustible Construction

Multiple Choice

Read each item carefully, then select the best response.

_______ **1.** Steel's compressive strength is __________ its tensile strength.

A. greater than
B. less than
C. equal to
D. not related to

_______ **2.** Steel tendons used for tensioned concrete can fail at what temperature?

A. 500°F
B. 700°F
C. 800°F
D. 900°F

_______ **3.** Atria roofs above __________ feet may be given an exemption from protection of the steel.

A. 40
B. 55
C. 75
D. 90

NOTES

_______ **4.** Water supplies for spray systems designed to protect steel supports for flammable liquid fires are calculated on a requirement of __________ gallon(s) per minute.

A. 0.25
B. 0.5
C. 0.75
D. 1

_______ **5.** Rigid frames can provide clear spans of about __________ feet.

A. 100
B. 125
C. 150
D. 175

_______ **6.** Which material is noncombustible but disintegrates rapidly in a fire?

A. Aluminum
B. Glass-fiber reinforced plastics
C. Cement-asbestos board
D. Steel

_______ **7.** All of the following are characteristics of steel, EXCEPT:

A. consistency of its structural characteristics.
B. strength.
C. inability to elongate.
D. ability to be connected to other structural elements.

_______ **8.** All of the following are types of girders, EXCEPT:

A. box girder.
B. lattice girder.
C. spandrel girder.
D. plate girder.

_______ **9.** Buildings in which the roof holds the tilt-up concrete wall panels in place are an example of which type of construction?

A. High-rise framing construction
B. Ordinary construction
C. Prefabricated construction
D. Tilt-slab construction

_______ **10.** Steel expands from 0.06 to 0.07 percent in length for each __________ rise in temperature.

A. 1°F
B. 10°F
C. 100°F
D. 1000°F

NOTES

Fill-in-the-Blank

Read each item carefully, then complete the statement by filling in the missing word(s).

1. Elongating steel exerts a/an __________ force against the structure that restrains it.
2. The better a building is tied together to resist wind load, the more likely it is to suffer __________ collapse due to fire distortion.
3. Interstitial space comes from a/an __________ term.
4. Cement-asbestos board is typically used for __________ construction.
5. Aluminum has a/an __________ melting point.
6. During a fire, the steel structure in a building should be kept __________.
7. The weight of steel sections is usually given per running __________.
8. Very deep parallel-chord trusses have been used as __________ beams in hospitals.
9. To achieve a/an __________ roof, the framing may consist of columns and beams with triangular trusses.
10. Because of the requirement for underground parking, building excavations are being made much __________ than in previous years.

True/False

Read each item carefully, then, if you believe the statement to be more true than false, write "T" in the space provided. If you believe the statement to be more false than true, write "F" in the space provided.

_____ 1. Non-combustible construction has moderate to heavy fire resistance.

_____ 2. The modulus of elasticity is high on steel.

_____ 3. Steel is a bad thermal conductor.

_____ 4. Today's high-rise construction commonly uses glass and metal panels.

_____ 5. Grease fires in ducts often extend to the structure by conduction.

_____ 6. Unprotected steel structures do not have a potential for early collapse.

_____ 7. A metal-deck roof fire can sometimes be fought from the underside.

_____ 8. Buildings with steel units tied together may undergo torsional or eccentric loads during a fire that exceed their capacity.

NOTES

_____ **9.** A small mass of steel can carry heavy loads.

_____ **10.** The method of wind bracing in a building is a serious concern to fire fighters after the building is completed.

Short Answer

Complete this section with short written answers, using the space provided.

1. What are the three classes of calculated risks outlined in the text?

____

____

____

2. Steel buildings can be broken down into four types of protection; what are they?

____

____

____

3. What methods can be used to protect steel?

____

____

____

4. What is the main difference between noncombustible and fire-resistive construction?

____

____

____

5. What are the benefits of using steel in building construction?

____

____

____

Word Fun

The following crossword puzzle is an activity provided to reinforce correct spelling and understanding of relevant terminology. Use the clues provided to complete the puzzle.

CLUES

Across

2 A standard I-beam cut lengthwise through the web

3 A piece of steel that has two legs at right angles

5 A steel structural component that has a square U-shaped cross section

6 Plates less than 6 inches in width

8 A column made of vertical units connected with diagonal pieces

Down

1 Flat pieces of steel

2 A type of beam used to laterally relocate the vertical load of columns to clear an opening area

4 Type of girder that will tie wall columns together in a framed building

7 A lightweight material that is both malleable and nonmagnetic

9 To evaluate and categorize

NOTES

Fire Alarms

The following case scenarios will give you an opportunity to explore the concerns associated with building construction for the fire service. Read each scenario, then answer each question in detail.

1. A building has been built of steel bar joists, metal roof deck, and masonry walls. A local businessperson is proposing to put a furniture store into the building. What concerns should be noted in the prefire plan related to the steel in this building?

____

____

____

____

____

2. A steel building is under construction when it catches fire on a windy day. What special considerations does this situation pose?

____

____

____

____

____

NOTES

Research Question

The following research question will give you an opportunity to explore the principles presented in *Brannigan's Building Construction for the Fire Service, Fourth Edition* in real situations within your community. Consider the following question and use the resources available to you to provide a detailed response on a separate piece of paper.

A fire in Charleston, South Carolina, killed nine fire fighters in June 2007. The building was a furniture store with a large warehouse. Conduct research on the available information regarding the fire and, using the information provided in the book, provide some theories or evidence from the media or other sources that may explain the reason for the collapse of the building.

Workbook Activities

The following activities have been designed to help you. Your instructor may require you to complete some or all of these activities as a regular part of your training program. You are encouraged to complete any activity that your instructor does not assign as a way to enhance your learning.

Matching

Match each of the terms in the left column to the appropriate definition in the right column.

_______ **1.** Aggregate

_______ **2.** Cast-in-place concrete

_______ **3.** Chairs

_______ **4.** Composite construction

_______ **5.** Falsework

_______ **6.** Footing

_______ **7.** Lally column

_______ **8.** Mudsill

_______ **9.** Reshoring

_______ **10.** Skewback

A. Buildings in which different load-bearing materials are used in different areas of the building

B. Temporary shoring, formwork, or lateral bracing to support concrete work in construction

C. Variety of materials added to a cement mixture to make concrete

D. Thick concrete pads that transfer loads of piers or columns to the ground

E. Concrete molded in the location where it is expected to remain

F. Shores that are put back into concrete to help carry the load of curing concrete

G. Devices designed to keep the rods up off the surface of the form

H. Tiles shaped to fit around steel

I. Planks on which formwork shores rest

J. Steel pipes filled with concrete

Chapter 10

Fire-Resistive Construction

Multiple Choice

Read each item carefully, then select the best response.

_______ 1. Which of the following is a characteristic of concrete?

A. Strong in compression

B. Weak in shear

C. No tensile strength

D. All of the above

_______ 2. Steel is significantly more __________ than concrete.

A. costly

B. prevalent

C. strong (compressive strength)

D. all of the above

_______ 3. High-tensile strength wire loses its pre-stress at what temperature?

A. 800°F

B. 850°F

C. 900°F

D. 1000°F

_______ 4. Brick veneer buildings are generally kept to a maximum of __________ stories.

A. 3

B. 4

C. 5

D. 6

NOTES

_______ **5.** Formwork can represent what percent of the cost of a concrete structure?
A. 0.06%
B. 0.6%
C. 6%
D. 60%

_______ **6.** The most dangerous fire potential in a concrete construction site is:
A. arson.
B. heating.
C. temporary electrical lines.
D. formwork.

_______ **7.** Telescoping tubular steel braces are called:
A. tormentors.
B. rakers.
C. brakers.
D. tendons.

_______ **8.** Compared to conventional reinforced concrete, post-tensioned concrete presents __________ catastrophic collapse hazard during a fire.
A. less
B. a greater
C. an equal
D. none of the above

_______ **9.** Which of the following provides diagonal bracing in precast buildings?
A. Columns
B. Wooden falsework
C. Cold-drawn steel cables
D. All of the above

_______ **10.** Which of the following provides protection for each piece of steel?
A. Membrane fireproofing
B. Fire rating
C. Individual fireproofing
D. None of the above

Fill-in-the-Blank

Read each item carefully, then complete the statement by filling in the missing word(s).

1. Cement formation is a __________ reaction.
2. In a fire, it is vital that any exposed steel connected to stressed tendons be __________ immediately.

NOTES

3. The value of listing building materials is contingent upon the __________ assembly being accomplished exactly as performed in the laboratory.

4. The disadvantage of concrete is its __________.

5. Thin rods installed near the surface of concrete to help concrete resist cracking due to temperature changes are called __________ rods.

6. In a concrete masonry structure, steel reinforcement is embedded in such a manner that the two materials act together in __________ forces.

7. A system that incorporates beams running in only one direction is called a/an __________ structural system.

8. Concrete floors in cast-in-place, concrete-framed buildings are cast integrally with __________, providing a monolithic rigid-framed building.

9. In __________ concrete, closely spaced beams are set at right angles to one another and unnecessary concrete is formed out.

10. The length of time required to cure concrete depends on the type of cement used and the __________ during the curing period.

True/False

Read each item carefully, then, if you believe the statement to be more true than false, write "T" in the space provided. If you believe the statement to be more false than true, write "F" in the space provided.

_____ 1. Cutting into a post-tensioned floor is unsafe during a fire.

_____ 2. Concrete has good shear resistance.

_____ 3. Brick veneer must have contraction provisions for high early-strength concrete.

_____ 4. Vertical reinforcing bars in concrete beams designed to prevent cracking under shear stresses are called stirrups.

_____ 5. The loads on concrete buildings under construction cannot exceed the design load.

_____ 6. Progressive collapse is not a concern after the concrete is hard to the touch.

_____ 7. A plenum ceiling requires ceiling panels to be in place for proper operation.

_____ 8. A monolithic concrete building is resistant to collapse.

_____ 9. If a building is concrete, the contents have very little effect on the building.

_____ 10. mitation concrete panels can be used on the exterior of a building.

NOTES

Short Answer

Complete this section with short written answers, using the space provided.

1. In what three areas do fire departments face problems regarding concrete construction?

____

____

____

2. What is the process of lift slab construction?

____

____

____

3. What is the purpose of steel reinforcing rods in columns?

____

____

____

4. What indicators show clear signs of possible trouble in a concrete building during a prefire plan?

____

____

____

5. What is the concern with impact loads on concrete?

____

____

____

Word Fun

The following crossword puzzle is an activity provided to reinforce correct spelling and understanding of relevant terminology. Use the clues provided to complete the puzzle.

CLUES

Across

1 Flow valve that controls a sudden increased flow and shuts off a flammable gas if a broken line occurs

4 Mold that shapes concrete

5 Often used to deliver materials to buildings under construction

6 Another name for a cable or tendon

7 Columns that use steel and concrete to form one unit

8 Placing fluid concrete into molds

9 Loss of surface material when concrete is subjected to heat

Down

2 A dangerous form of concrete pouring that involves moving forms upward as the concrete is poured

3 Type of construction in which one pour of concrete is completed in order to bond all of the concrete together

4 Insulating steel

NOTES

Fire Alarms

The following case scenarios will give you an opportunity to explore the concerns associated with building construction for the fire service. Read each scenario, then answer each question in detail.

1. A local contractor is building a cast-in-place concrete structure that is five stories tall. What tactical hazards should be considered as this building is being built?

2. A building with a floor-ceiling assembly is on fire. What factors might cause problems for fire fighters by compromising the fire performance of the assembly?

NOTES

Research Question

The following research question will give you an opportunity to explore the principles presented in *Brannigan's Building Construction for the Fire Service, Fourth Edition* in real situations within your community. Consider the following question and use the resources available to you to provide a detailed response on a separate piece of paper.

Contact a local cement company and find out the cure times that are normal for concrete of different types of pours. Ask the concrete company about the methods that may be most effective when removing concrete during an emergency. (Many times they have experienced the removal of concrete from the inside of a broken-down cement delivery truck.)

Workbook Activities

The following activities have been designed to help you. Your instructor may require you to complete some or all of these activities as a regular part of your training program. You are encouraged to complete any activity that your instructor does not assign as a way to enhance your learning.

Matching

Match each of the terms in the left column to the appropriate definition in the right column.

_______ 1. Calcination

_______ 2. Control area

_______ 3. Friable

_______ 4. Occupancy

_______ 5. Passive fire protection

_______ 6. Pause

_______ 7. Projected beam

_______ 8. Smoke barrier

_______ 9. Stack effect

_______10. Turned mass dampers

A. Easily disintegrated

B. Intended use of a building

C. A building or portion of a building within which hazardous materials are allowed to be stored in quantities not exceeding the maximum allowable quantities

D. Vertical airflow within buildings caused by the temperature-created density differences between the building interior and exterior

E. Deterioration of a product by heating it to high temperatures

F. Heavy weights installed high up in a building that are adjusted by computers

G. Detector that covers large areas with a beam of light

H. Continuous membrane to restrict the spread of smoke

I. Material applied to substrate and designed to protect it from thermal effects

J. In reference to atmospheric conditions, the layer of air warmer than the air below

Chapter 11

Specific Occupancy Details and Hazards

Multiple Choice

Read each item carefully, then select the best response.

_______ 1. NIST stands for what?

A. National Institute for Safety and Technology
B. National Institute for Standards and Technology
C. National Institute for Standards and Tactics
D. Nationwide Institute for Safety and Techniques

_______ 2. Combustible multiple dwellings include:

A. garden apartments.
B. modern row houses.
C. town houses.
D. all of the above.

_______ 3. Preincident plan review should make sure that there is __________ of clear width around the building.

A. ½ foot
B. 2 feet
C. 20 feet
D. 200 feet

NOTES

_______ **4.** All of the following are types of exterior structural walls of typical garden apartments, EXCEPT:

A. solid masonry.

B. brick veneer over platform steel frame.

C. partially solid masonry.

D. wood.

_______ **5.** The greatest flow in a "winter stack effect" will be at the __________ floor.

A. first

B. top

C. middle

D. largest

_______ **6.** In a summer stack effect, the flow is:

A. upward.

B. equal across floors except in high heat conditions.

C. slower.

D. downward.

_______ **7.** Draftstopping is usually required to limit the size of attic compartments to what amount?

A. 3 square feet

B. 30 square feet

C. 300 square feet

D. 3000 square feet

_______ **8.** Which of the following construction materials does not yield heat when burned in pure oxygen?

A. Wood

B. Steel

C. Gypsum

D. All of the above

_______ **9.** Which of the following requirements do building codes regulate for atria?

A. Full sprinkler protection

B. Smoke control system

C. Standby power

D. All of the above

_______ **10.** The most dangerous time in terms of a fire at places of worship is:

A. before initial construction is complete.

B. during renovations.

C. after renovations are complete.

D. 25 or more years after initial construction.

NOTES

Fill-in-the-Blank

Read each item carefully, then complete the statement by filling in the missing word(s).

1. Leasing plans showing __________ locations are required for fire protection in malls.
2. Brick or stone __________ serves as a heat sink for winter warmth.
3. In older parts of many cities, frame buildings were often erected in __________.
4. Under __________ conditions, smoke will move up and away for a fire.
5. Stack effect is __________ to the difference between the outside and inside temperatures.
6. A/An __________ stairway is designed to keep a stairway free of smoke.
7. A house that has a steep-pitched roof and is usually 1½ stories is called a/an __________.
8. __________-style homes are characterized by significant amounts of ornamentation, steep-pitched roofs, and a balloon-frame construction.
9. In a/an __________ bungalow-style home, there is often no ridge beam.
10. One-story __________ houses may fail to have a smoke detector in the attic.

True/False

Read each item carefully, then, if you believe the statement to be more true than false, write "T" in the space provided. If you believe the statement to be more false than true, write "F" in the space provided.

_____ 1. Garden apartments many times have plumbing walls aligned.

_____ 2. Specific hazards that affect the design and construction of a building are often regulated by building and fire codes.

_____ 3. There are several types of rated ceilings.

_____ 4. Cementing nail heads is part of a fire-resistive system.

_____ 5. Usually, a manufacturer will add some degree of safety beyond the test results in a fire-resistive testing.

_____ 6. Unprotected openings are frequently cut into firewalls.

_____ 7. Firewalls always extend through the roof.

_____ 8. Class I standpipe systems in a mall typically have self-sufficient pressure.

NOTES

_____ **9.** In a modern office building, as much as 25 percent of the floor volume may be in ceiling voids.

_____ **10.** Model building codes have allowed as much as 75 percent of exit stairwells to terminate into a building's lobby.

Short Answer

Complete this section with short written answers, using the space provided.

1. What problems do elevators pose for fire fighters?

____

____

____

2. What is the easiest way to get from a corridor into an apartment?

____

____

____

3. Elevator door restrictors are used to safeguard against what hazard?

____

____

____

4. What is the difference in wind on the lower floors of a building compared to the first floor?

____

____

____

5. What act encouraged improvements in fire safety nationwide in temporary lodgings?

____

____

____

Word Fun

The following crossword puzzle is an activity provided to reinforce correct spelling and understanding of relevant terminology. Use the clues provided to complete the puzzle.

CLUES

Across

2 An area 18 inches above or below the landing floor

6 Type of gas that ignites in air without the introduction of an ignition source

9 Filling the area between studs with mortar

10 A characteristic that means incapable of self-preservation

Down

1 Pallet storage in which pallets are not stacked with product

3 Hallways or tunnels used as an exit component and separated from other parts of the building

4 The large, ornamental opening and wall that separates an audience from a stage

5 A vertical stair with a width of not more than 24 inches

7 Any combination of vertical, horizontal, and diagonal members that supports stored materials

8 Type of smoke that falls downward

NOTES

Fire Alarms

The following case scenarios will give you an opportunity to explore the concerns associated with building construction for the fire service. Read each scenario, then answer each question in detail.

1. You have been asked to conduct a class for nursing home staff in case of a fire. What procedures should you teach the staff?

2. A company has decided to build a large warehouse in town; it will be a regional distribution center for a "big box" hardware and lumber store. What concerns might you have about this building?

NOTES

Research Question

The following research question will give you an opportunity to explore the principles presented in *Brannigan's Building Construction for the Fire Service, Fourth Edition* in real situations within your community. Consider the following question and use the resources available to you to provide a detailed response on a separate piece of paper.

Identify several elevators in your primary response area. List how many elevator banks there are, the types of lifting mechanisms, the locations of the equipment rooms, and any other special hazards associated with them. Would you use them in case of a fire? Can you identify ways to release a stuck elevator?

Workbook Activities

The following activities have been designed to help you. Your instructor may require you to complete some or all of these activities as a regular part of your training program. You are encouraged to complete any activity that your instructor does not assign as a way to enhance your learning.

Matching

Match each of the terms in the left column to the appropriate definition in the right column.

_______ 1. Global
_______ 2. Partial
_______ 3. Progressive
_______ 4. Secondary
_______ 5. Pancake
_______ 6. Lean-to floor
_______ 7. V-shaped
_______ 8. Cantilever
_______ 9. Tent
_______ 10. Lean over

A. Type of collapse that occurs after an initial collapse

B. Type of collapse in which the floor of a building collapses and one end of the floor is still supported, creating a triangular void space

C. Type of collapse involving the complete failure of the building

D. Type of collapse in which a relatively small area of structural damage results in a loss of localized load-carrying capability

E. Type of collapse in which one end of the collapsed floor is supported by an interior wall, creating two void spaces

F. Type of collapse in which the floor falls near its center, with the perimeter of the floor still partially supported by the walls

G. Type of collapse in which one end of the floor is still supported while the other is not

H. Type of collapse resulting from a failure of the portion of the building

I. Type of collapse limited to wood-frame structures

J. Type of collapse in which the building's floors stack on top of each other

Chapter 12

Collapse

Multiple Choice

Read each item carefully, then select the best response.

_______ 1. Which of the following has the greatest potential to kill large numbers of fire fighters in one incident?

A. Heart attack

B. Collapse

C. Smoke inhalation

D. Falls

_______ 2. Who is responsible for anticipating a collapse?

A. Incident commander

B. Operations chief

C. Safety officer

D. All of the above

_______ 3. In which of the following types of collapse do the walls fall straight down?

A. Curtain fall wall collapse

B. Inward outward collapse

C. Lean-to floor collapse

D. 90-degree wall collapse

NOTES

_______ 4. In which of the following types of collapse do the exterior walls fall horizontally?
   A. Curtain fall wall collapse
   B. Inward outward collapse
   C. Lean-to floor collapse
   D. 90-degree wall collapse

_______ 5. Collapse zones are generally what distance from a wall?
   A. One-third the height of the wall
   B. One-half the height of the wall
   C. The height of the wall
   D. Twice the height of the wall

Labeling

Label each type of collapse.

1. ______________________________

NOTES

2. ____

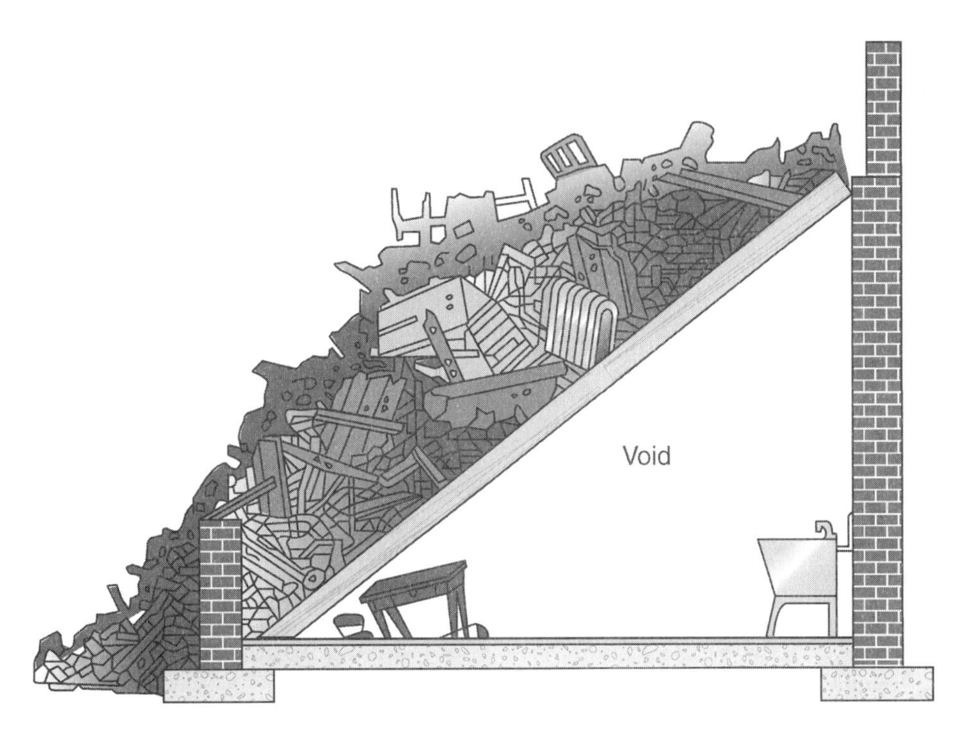

3. ____

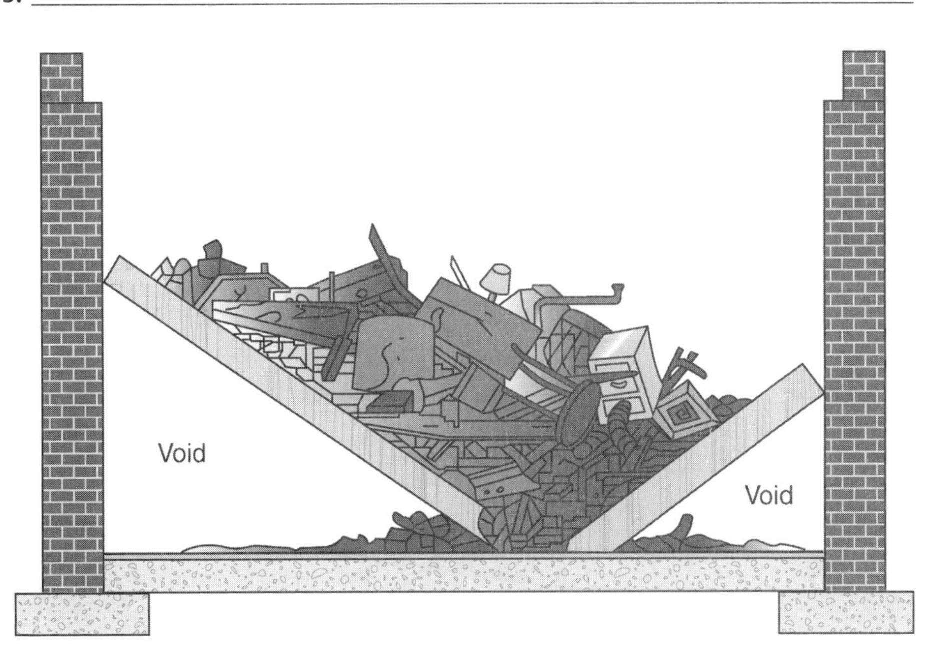

NOTES

4. ______________________________

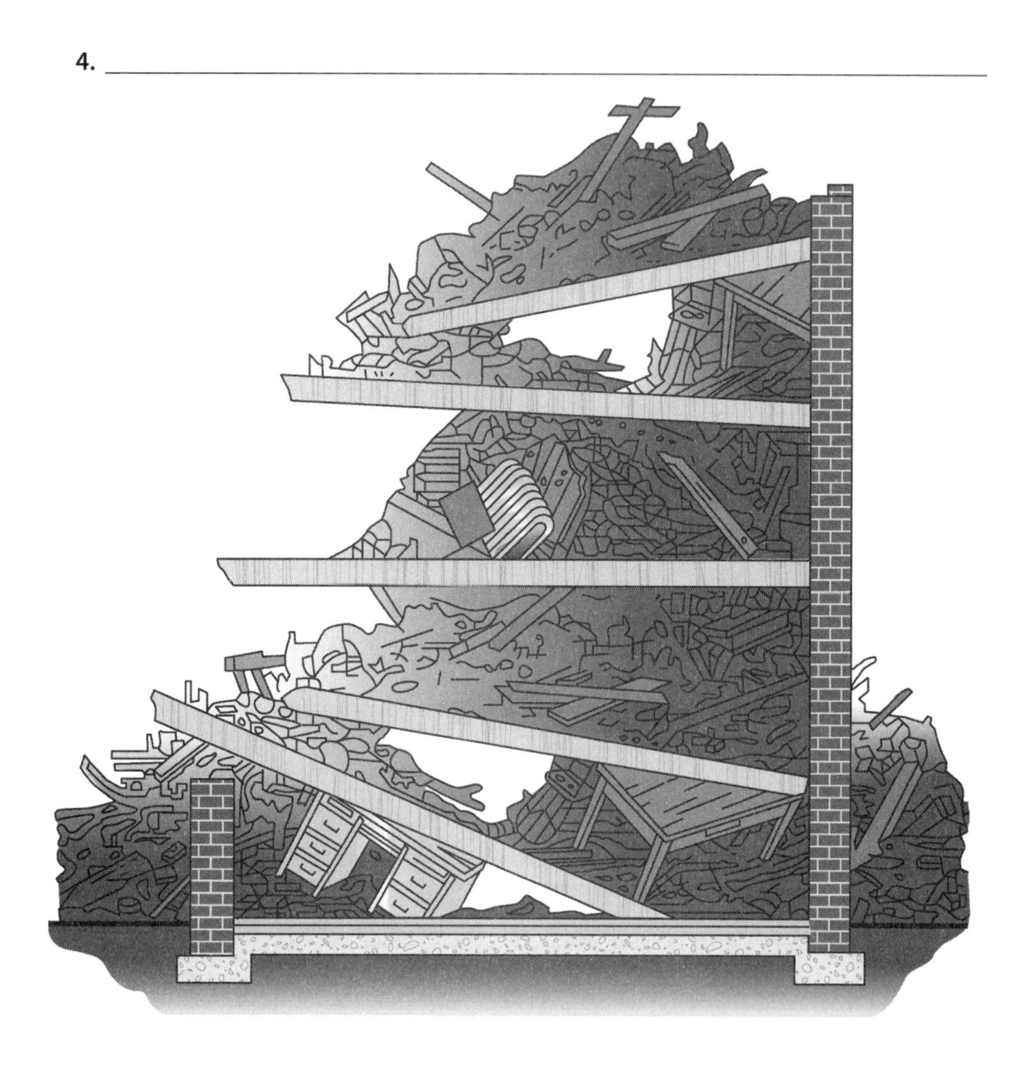

5. ______________________________

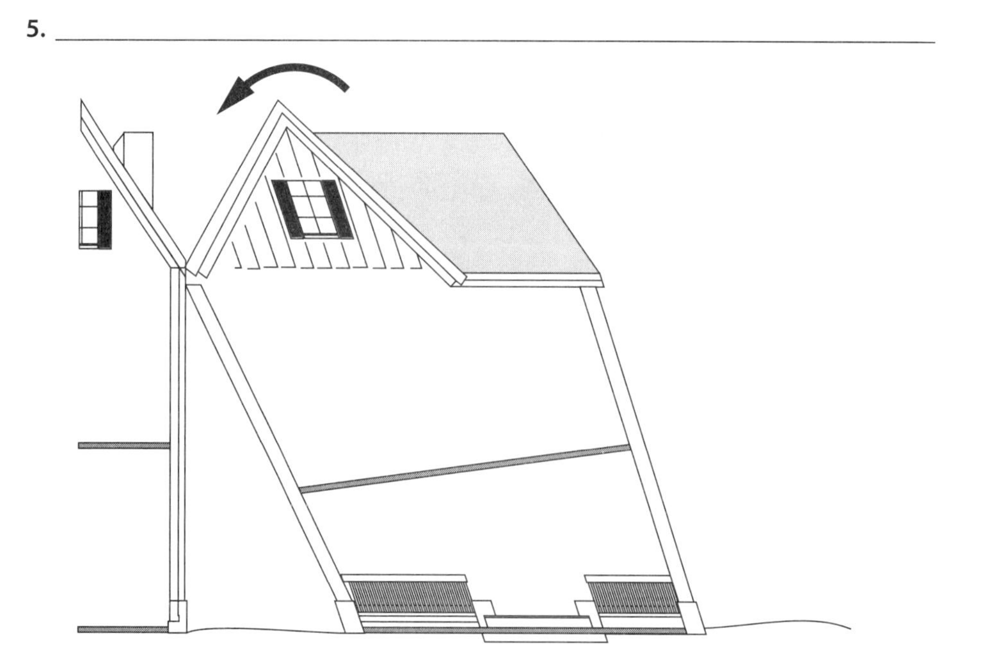

NOTES

6. ____

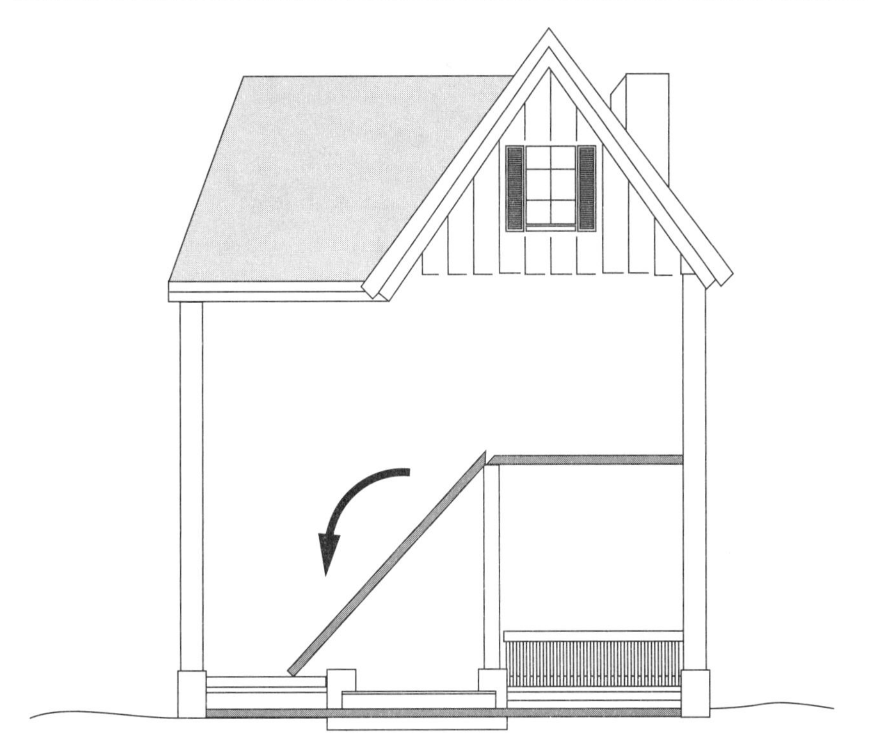

7. ____

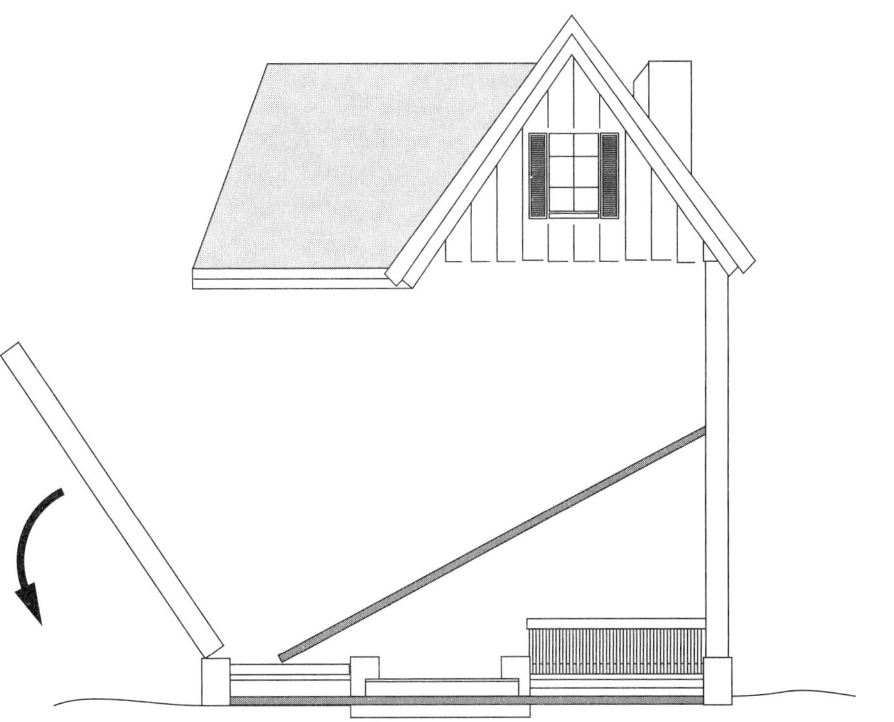

NOTES

8. ______________________________

9. ______________________________

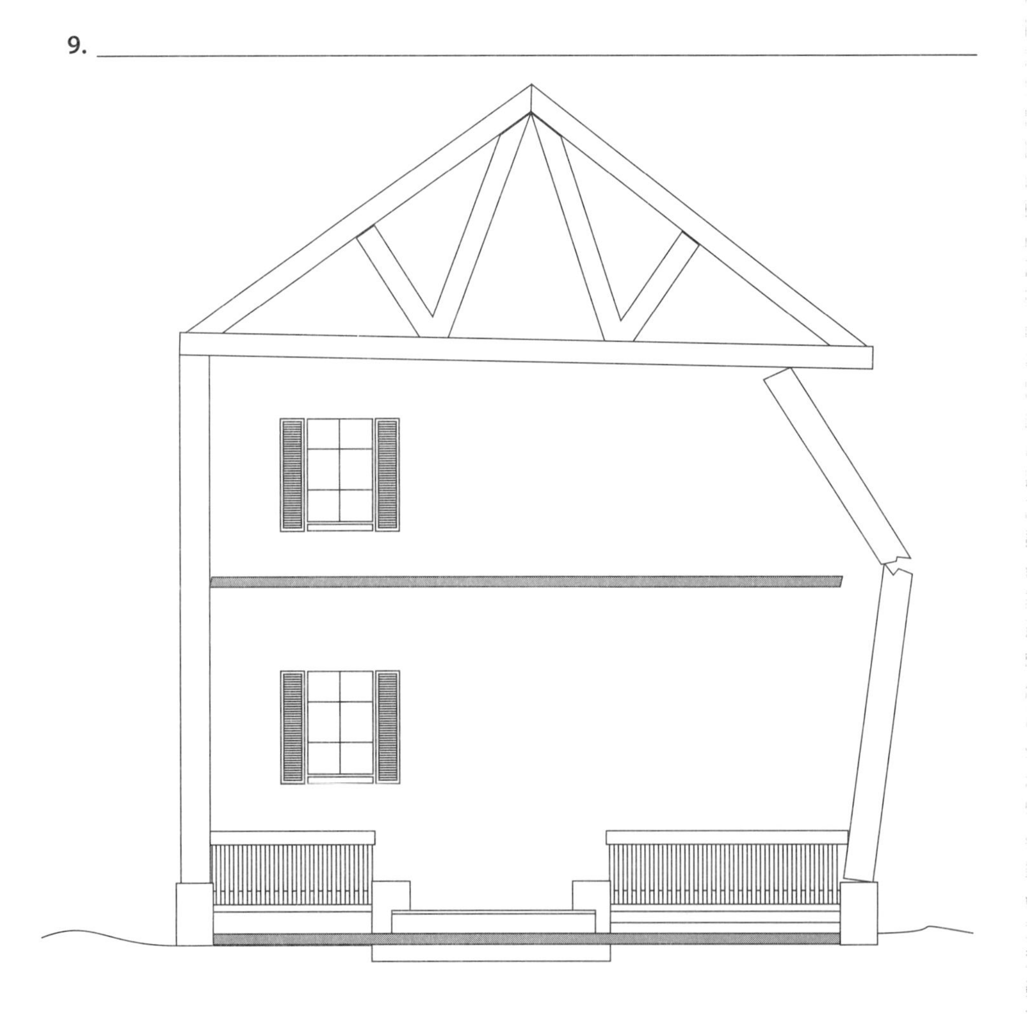

NOTES

Research Question

The following research question will give you an opportunity to explore the principles presented in *Brannigan's Building Construction for the Fire Service, Fourth Edition* in real situations within your community. Consider the following question and use the resources available to you to provide a detailed response on a separate piece of paper.

Using *Brannigan's Building Construction for the Fire Service, Fourth Edition; NFPA 1620: Recommended Practice for Pre-Incident Planning, 2003 Edition* (available at http://www.nfpa.org/freecodes/free_access_document.asp); and any other resources pertaining to building construction and fire fighter safety, conduct research on the history of one of the following buildings and write a pre-fire plan:

- A warehouse of at least 50,000 square feet
- A multi-family residential or mixed use building of at least 15,000 square feet per floor
- A high rise of any use group
- An industrial-type building with known hazards, such as storage or manufacture of chemicals
- A building in commercial use that is at least 75 years old and 5000 square feet (this may include several occupancies in a downtown area that are connected to each other)

Answer Key

Chapter 1: Introduction

Matching

1. C (page 5)
2. D (page 5)
3. A (page 5)
4. B (page 5)
5. E (page 5)
6. F (page 6)

Multiple Choice

1. D (page 5)
2. B (page 4)
3. B (page 6)

Short Answer

1. All of the following: (pages 4–7)
   - Type of construction
   - Present conditions
   - Type of operation
   - Command identification
   - Location of command
   - Any other pertinent information
2. Any three of the following: (pages 4–7)
   - An appropriate level of detail (not too specific or too detailed)
   - Separate inspections and surveys
   - Efforts to improve community cooperation
   - Single point-person to communicate with business owners
   - Consideration of potential emergencies
   - Plans for dealing with possible hazards
   - Conclusion of disaster potential

Chapter 2: Concepts of Construction

Matching

1. F (page 13)
2. C (page 13)
3. I (page 13)
4. A (page 13)
5. B (page 13)
6. D (page 13)
7. J (page 15)
8. G (page 15)
9. H (page 19)
10. E (page 16)

Multiple Choice

1. B (page 30)
2. C (page 35)
3. A (page 24)
4. D (page 23)
5. A (page 26)
6. A (page 26)
7. B (page 26)
8. A (page 32)
9. C (page 36)
10. D (page 37)

Labeling

(page 23)

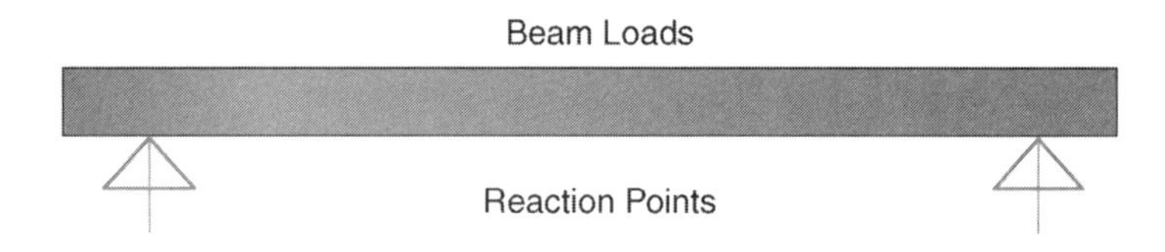

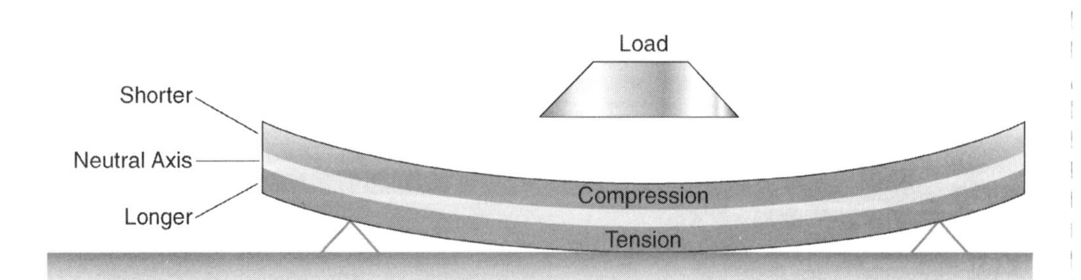

Fill-in-the-Blank

1. Glitch plate girder (page 22)
2. composite (page 22)
3. tube construction (page 18)
4. heat release rate (page 19)
5. needle beam (page 25)
6. beam (page 23)
7. nonvertical (page 26)
8. transfer (page 26)
9. transmitted (page 33)
10. plastic (page 34)

True/False

1. F (page 18)
2. T (page 18)
3. F (page 21)
4. T (page 20)
5. F (page 21)
6. T (page 18)
7. F (page 20)
8. F (page 24)
9. F (page 25)
10. T (page 27)

Short Answer

1. Panel walls (also called curtain walls) are non-load-bearing walls; party walls, however, are load-bearing walls. (page 30)
2. Pinned (when the elements are connected by simple connectors such as bolts, rivets, or welded joints) and rigid-framed (when the connections are strong enough to reroute forces if a member is removed). (page 34)
3. Any of the following: (page 35)
   - Masonry walls can shift outward, dropping joists.
   - Steel connectors can rust, eventually failing.
   - Concrete can disintegrate into serious failure.
   - Temporary field bolting can give way in high winds.
   - Loads can shift, causing a break in the connector.
4. Any three of the following: (page 31)
   - Buttresses
   - Pilasters
   - Wall columns
   - Cavity or hollow walls
5. Very long, thin columns (page 28)

Word Fun

Across

3. Voussoir (page 33)

6. Headers (page 29)

7. Buttress (page 31)

8. Raker (page 26)

10. Pilaster (page 31)

Down

1. Chord (page 22)

2. Firecut (page 36)

4. Vierendeel (page 18)

5. Arch (page 32)

9. Demising (page 30)

Chapter 3: Methods and Materials of Construction, Renovation, and Demolition

Matching

1. D (page 47)
2. I (page 54)
3. J (page 48)
4. F (page 58)
5. G (page 50)
6. A (page 58)
7. B (page 57)
8. H (page 56)
9. C (page 56)
10. E (page 54)

Multiple Choice

1. A (page 52)
2. C (page 53)
3. C (page 55)
4. B (page 52)
5. C (page 52)
6. A (page 48)
7. D (page 49)
8. C (page 47)
9. C (page 47)
10. D (page 49)

Fill-in-the-Blank

1. building (page 47)
2. fire (page 47)
3. Zoning (page 47)
4. Americans with Disabilities Act (page 47)
5. Occupational Safety and Health Administration (page 47)
6. architect (page 47)
7. engineer (page 47)
8. civil (page 47)
9. mechanical (page 47)
10. electrical (page 47)

True/False

1. T (page 49)
2. F (page 47)
3. T (page 49)
4. T (page 50)
5. F (page 52)
6. F (page 54)
7. F (page 62)
8. T (page 58)
9. T (page 52)
10. T (page 52)

Short Answer

1. Hose valves may not be capped, water supply may not be to all floors, or there may be debris in system. (page 50)
2. Zoning regulations (page 47)
3. Americans with Disabilities Act (page 47)
4. Any of the following: (page 47)
   - Architect
   - Structural engineer
   - Civil engineer
   - Mechanical engineer
   - Electrical engineer
   - Fire protection engineer
   - General contractor
   - Subcontractor
   - Electrical contractor
   - Plumbing contractor
   - Wallboard contractor
   - Fire alarm/security system contractor
   - Sprinkler contractor
   - Fireproofing contractor
5. Steel conducts heat, elongates and may push through barriers, and fails at about 1000°F. (page 55)

Fire Alarms

1. Special tools may be necessary to cut through these windows. (page 55)
2. Concrete that is not cured may be subject to collapse. One of the worst collapses that can occur is a progressive collapse. (page 50)

Chapter 4: Building and Fire Codes

Matching

1. D (page 72)
2. H (page 72)
3. F (page 72)
4. A (page 75)
5. C (page 75)
6. E (page 74)
7. G (page 74)
8. B (page 74)
9. J (page 67)
10. I (page 67)

Multiple Choice

1. B (page 72)
2. A (page 74)
3. D (page 70)
4. D (page 71)
5. B (page 68)
6. D (page 68)
7. B (page 68)
8. B (page 67)
9. C (page 67)
10. B (page 74)

Fill-in-the-Blank

1. gross (page 73)
2. trade-offs (page 70)
3. Assembly Occupancy
   > 1000 occupants (page 69)
4. Assembly "A-3" (page 69)
5. legacy (page 67)
6. ASTM E-84 (page 67)
7. subcategory (page 67)
8. Hybrid (page 68)
9. prohibited (page 74)
10. index (page 74)

True/False

1. T (page 72)
2. F (page 72)
3. F (page 73)
4. T (page 74)
5. T (page 68)
6. T (page 69)
7. T (page 66)
8. F (page 67)
9. T (page 67)
10. F (page 67)

Short Answer

1. (page 67)
   - Type I: Fire resistive
   - Type II: Noncombustible
   - Type III: Ordinary
   - Type IV: Heavy Timber
   - Type V: Wood Frame
2. Electric cable and hydraulic piston (page 75)
3. Any of the following: (page 72)
   - Number of exit stairwells
   - Travel distance
   - Width of a doorway
4. Any of the following: (pages 72–74)
   - Occupant load
   - Number of exit paths or doors for each room
   - Number of exit paths or doors for the building
   - Rated corridors
   - Door hardware
   - Exterior emergency escape and rescue windows
   - Burglar bars
5. Elevator door restrictors are metal bars used to prevent the occupants of an elevator stuck between floors from opening the car and shaft doors and falling down the elevator shaft below. (page 75)

Fire Alarms

1. The floors supporting more than one floor or one floor should be 2 hours. The floor supporting only the roof should be 1 hour. (page 68)
2. The number of stories is 4, and the area per floor is 26,000 square feet. (page 71)

Chapter 5: Features of Fire Protection

Matching

1. B (page 121)
2. J (page 84)
3. D (page 100)
4. H (page 121)
5. C (page 106)
6. G (page 119)
7. E (page 93)
8. I (page 109)
9. F (page 82)
10. A (page 84)

Multiple Choice

1. C (page 81)
2. A (page 119)
3. B (page 118)
4. D (page 121)
5. B (page 118)
6. C (page 119)
7. D (page 106)
8. A (page 98)
9. C (page 97)
10. A (page 95)

Fill-in-the-Blank

1. rated (page 81)
2. specified (page 93)
3. Habel's (page 97)
4. self-closing (page 100)
5. open (page 114)
6. Aerosol (page 117)
7. control (page 121)
8. fusible (page 100)
9. blocked (page 102)
10. two (page 102)

True/False

1. T (page 81)
2. F (page 81)
3. T (page 81)
4. T (page 81)
5. T (page 83)
6. T (page 82)
7. F (page 84)
8. T (page 85)
9. F (page 88)
10. F (page 92)

Short Answer

1. Interior finish increases fire extension by surface flame spread, generates smoke and hot gases, and adds fuel to the fire, contributing to flashover. (page 85)

2. Any of the following: (page 80)
   - Restrictions on the area/height of a building
   - Limits on the combustibility of roofs and exterior wall surfaces
   - Minimum separation distances between buildings
   - Limits on openings in exterior walls

3. Any of the following: (page 80)
   - Fire-rated floors
   - Protection of vertical floor openings
   - Compartmentation
   - Fire-resistive construction
   - Use of fire protection systems

4. Any of the following: (page 81)
   - Floors or roofs softening
   - Water flowing through bricks
   - Smoke pushing out of mortar joints
   - Strange noises

5. Cycling sprinklers are designed to reduce water damage and permit lower water requirements. (page 117)

Word Fun

Across

3. Batt (page 84)

4. Purge (page 121)

7. Clean agent (page 122)

8. Carbon dioxide (page 122)

9. Compartmentation (page 80)

Down

1. Combustible acoustical tile (page 85)

2. Backdraft (page 82)

3. Bagasse (page 85)

5. Silence (page 121)

6. Flameproof (page 81)

Fire Alarms

1. Either of the following: (pages 102, 116, and 117)
   - The sprinklers may not operate correctly or may be impaired by blockage.
   - Sprinklers may not be adequate for the stored items within the space and would have no effect on smoke or gases that may go through the openings of the escalator.

2. Pre-planning considerations include: (pages 121–122)
   - A general description of the system
   - Location and extent of the system
   - System design criteria
   - If it is automatic, manual, or both
   - What types of initiating devices activate the system
   - Step-by-step operating sequence
   - Location and description of control panel

   Firefighting considerations include: (page 122)
   - Assess the system operation: Is it working properly? If not, should you shut it down?
   - When activating a system manually, let all fire fighters know so they will not be endangered.
   - Leave a fire fighter at the control panel with a radio to ensure quick activation/shutdown if necessary.

Chapter 6: Wood Frame Construction

Matching

1. B (page 135)
2. E (page 137)
3. H (page 134)
4. C (page 134)
5. A (page 131)
6. F (page 132)
7. D (page 130)
8. I (page 130)
9. J (page 130)
10. G (page 130)

Multiple Choice

1. D (page 137)
2. B (page 135)
3. C (page 136)
4. C (page 138)
5. A (page 140)
6. D (page 143)
7. B (page 144)
8. C (page 146)
9. C (page 145)
10. D (page 132)

Fill-in-the-Blank

1. foundation (page 132)
2. curtain (page 130)
3. Collapse (page 133)
4. dead (page 137)
5. sawn (page 139)
6. horizontal (page 141)
7. vertical (page 142)
8. Glulam (page 143)
9. Finger (page 144)
10. four (page 145)

True/False

1. F (page 144)
2. T (page 146)
3. F (page 144)
4. T (page 140)
5. T (page 138)
6. F (page 138)
7. F (page 136)
8. T (page 137)
9. F (page 135)
10. T (page 132)

Short Answer

1. Inverted king and queen trusses (page 137)
2. Struts, ties, and panel points (page 137)
3. Inherent and legal (page 141)
4. Plywood may delaminate when exposed to fire, which increases the surface area and its rate of heat release. (page 143)
5. Roofing materials are rated as Class A–C and are rated for flame exposure, spread of flame, and resistance to burning brands or flying pieces of burning wood. (page 147)

Word Fun

Across

4. Vermiculite (page 134)

5. Column (page 138)

7. Ridge (page 135)

8. Tie (page 137)

Down

1. Connector (page 137)

2. Panel (page 137)

3. Hip (page 135)

4. Valley (page 135)

6. Trimmer (page 135)

8. Trunnel (page 132)

Fire Alarms

1. Restaurants of wood construction usually include trusses in their roof structures. Potential hazards include large areas of voids, large dead loads, hidden fires, and quicker collapse. (pages 136–141)

2. Houses built around that time generally were balloon-frame. This house may not have any firestopping. Firefighters should anticipate that the wood may be damaged in some areas and that there may be voids. Additionally, sprinklers are probably not present or may not be effective. Basement fires may allow the fire to travel through the building quickly. (pages 132–133)

Chapter 7: Heavy Timber and Mill Construction

Matching

1. H (page 157)
2. G (page 159)
3. B (page 157)
4. A (page 157)
5. F (page 158)
6. D (page 156)
7. E (page 157)
8. I (page 156)
9. J (page 157)
10. C (page 158)

Multiple Choice

1. A (page 156)
2. C (page 157)
3. D (page 157)
4. B (page 158)
5. C (page 159)

True/False

1. T (page 157)
2. F (page 157)
3. F (page 159)
4. T (page 161)
5. F (page 159)

Short Answer

1. The slow-burning characteristic is an advantage only as long as the fire department can maintain interior offensive operations. (page 157)
2. Vacant buildings often do not have adequate or functional sprinkler systems. They may also have additional structural hazards or conditions. (page 161)

Chapter 8: Ordinary Construction

Matching

1. E (page 186)
2. C (page 169)
3. A (page 187)
4. H (page 169)
5. B (page 181)
6. J (page 167)
7. D (page 169)
8. F (page 181)
9. G (page 181)
10. I (page 169)

Multiple Choice

1. B (page 166)
2. B (page 188)
3. C (page 180)
4. A (page 181)
5. D (page 178)
6. C (page 178)
7. D (page 177)
8. C (page 167)
9. C (page 167)
10. D (page 172)

Fill-in-the-Blank

1. parging (page 174)
2. arch (page 176)
3. walls (page 176)
4. Carbon monoxide (page 178)
5. lateral (page 169)
6. outward (page 184)
7. cornice (page 188)
8. Party (page 189)
9. prefire (page 189)
10. blocking (page 190)

True/False

1. T (page 167)
2. T (page 167)
3. F (page 167)
4. F (page 168)
5. T (page 170)
6. T (page 172)
7. T (page 173)
8. T (page 174)
9. F (page 175)
10. T (page 175)

Short Answer

1. Any of the following: (pages 174–178)
    - Bricks and mortar
    - Wood beams
    - Cracks
    - Arches
    - Wall weakness
    - Steel lintels
    - Bracing
    - Eccentric loads
    - Unvented voids
    - Planes of weakness
2. Arches and beams (page 173)
3. Any of the following: (page 167)
    - Brick
    - Steel
    - Wood
    - Stone
    - Concrete block
    - Cast iron
4. Any of the following: (page 177)
    - Stars
    - Plates
    - Channel sections
    - Spreaders or straps tying the wall to the side wall
5. Any of the following: (page 177)
    - Projecting sign
    - Air conditioner
    - Balcony

Word Fun

Across

5. Tied arch (page 184)

6. Veneer (page 169)

7. Rubble (page 169)

8. Transfer (page 181)

10. Flying buttress (page 169)

Down

1. Cross wall (page 169)

2. Wythe (page 169)

3. Light well (page 182)

4. Stretcher course (page 169)

9. Fire limit (page 167)

Fire Alarms

1. Any of the following: (page 191)
   - Available fire load
   - Type of construction
   - Nature of connections to adjacent buildings
   - Water supplies
   - Protection provided
   - Damaged areas of the building
   - Hidden voids
   - Suspended or other loads

2. General indicators include: (page 174)
   - Inherent structural instability, aggravated by fire
   - Failure of a nonmasonry supporting element upon which some of the masonry depends
   - Increase in the live load due to fire fighting operations, specifically retained water
   - Collapse of floor or roof with consequent impact load to the masonry wall
   - Impact load of an explosion
   - Collapse of a masonry unit due to overheating
   - Collapse of another building onto the building in question
   - Masonry cracks occurring or widening

Chapter 9: Noncombustible Construction

Matching

1. J (page 202)
2. E (page 201)
3. F (page 206)
4. C (page 203)
5. B (page 211)
6. A (page 200)
7. H (page 209)
8. I (page 201)
9. G (page 201)
10. D (page 209)

Multiple Choice

1. C (page 198)
2. C (page 199)
3. B (page 210)
4. A (page 215)
5. A (page 202)
6. A (page 203)
7. C (page 201)
8. B (page 201)
9. D (page 205)
10. C (page 207)

Fill-in-the-Blank

1. lateral (page 206)
2. progressive (page 203)
3. medical (page 203)
4. friable (page 203)
5. low (page 203)
6. cool (page 199)
7. foot (page 201)
8. floor (page 202)
9. peaked (page 202)
10. deeper (page 209)

True/False

1. F (page 198)
2. T (page 198)
3. F (page 199)
4. T (page 205)
5. T (page 206)
6. F (page 215)
7. T (page 214)
8. T (page 202)
9. T (page 198)
10. F (page 205)

Short Answer

1. All of the following: (page 211)
   - Financial or economic
   - Engineering
   - Forget it
2. All of the following: (page 215)
   - Unprotected
   - Dynamically protected
   - Passively protected
   - Passive/dynamic combination protection
3. Membrane protection and sprayed-on fireproofing (page 216)
4. The level of fire resistance (fire rating) assigned to the structural frame, walls, floors, and roof (page 198)
5. Any of the following: (page 198)
   - Strength
   - Wide availability
   - Inexpensive cost

Word Fun

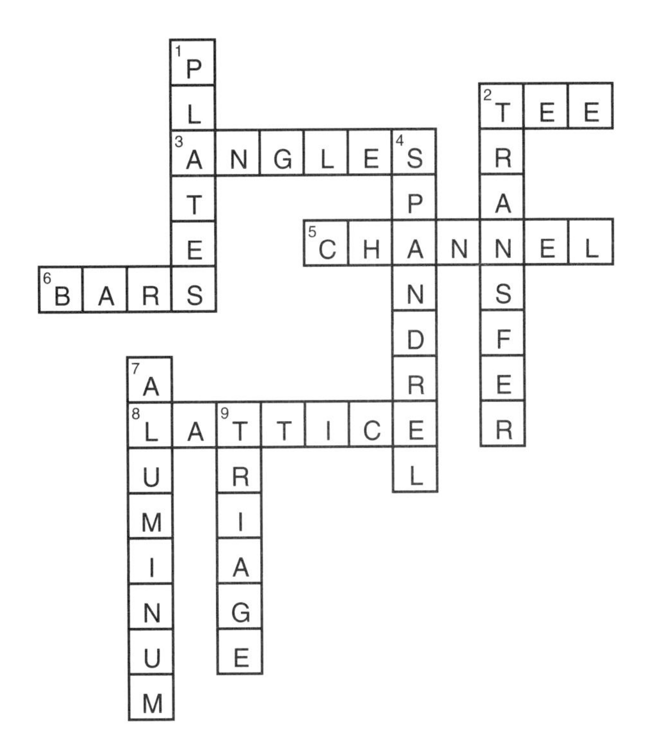

Across

2. Tee (page 200)

3. Angles (page 200)

5. Channel (page 200)

6. Bars (page 200)

8. Lattice (page 200)

Down

1. Plates (page 200)

2. Transfer (page 203)

4. Spandrel (page 200)

7. Aluminum (page 203)

9. Triage (page 189)

Fire Alarms

1. The contents of the building can dramatically affect the structure. It should be noted how the underside of the roof could be reached with hose streams to minimize heat damage to the joists. Special attention should be directed at the building content height as well as the general height of the building. These factors could influence the amount of time it takes for the building to collapse or become heavily involved in a fire that prevents the steel from being cooled by heavy streams. (pages 210–212)

2. The temporary bracing should be examined closely. It is impossible to make all the connections as soon as steel is placed. Temporary bolts are placed in rivet holes; thereafter, rivets are driven in the remaining holes, then the bolts are replaced with rivets, or permanent bolts. This practice is known as "field bolting," and it is good to give a wide berth to field-bolted structures during high winds. (page 200)

Chapter 10: Fire-Resistive Construction

Matching

1. C (page 226)
2. E (page 225)
3. G (page 226)
4. A (page 226)
5. B (page 233)
6. D (page 226)
7. J (page 227)
8. I (page 234)
9. F (page 234)
10. H (page 240)

Multiple Choice

1. A (page 229)
2. C (page 229)
3. A (page 231)
4. D (page 232)
5. D (page 233)
6. B (page 235)
7. A (page 237)
8. B (page 236)
9. C (page 237)
10. C (page 241)

Fill-in-the-Blank

1. chemical (page 224)
2. cooled (page 237)
3. field (page 238)
4. weight (page 240)
5. temperature (page 228)
6. resisting (page 228)
7. one-way (page 227)
8. columns (page 231)
9. waffle (page 230)
10. temperature (page 234)

True/False

1. T (page 237)
2. F (page 225)
3. T (page 225)
4. T (page 229)
5. F (page 233)
6. F (page 235)
7. T (page 238)
8. T (page 245)
9. F (page 242)
10. T (page 243)

Short Answer

1. All of the following: (page 225)
   - Collapse during construction with no fire
   - Fire during construction
   - Fire in completed, occupied buildings
2. Columns are erected to their full height and then each floor is poured and lifted into place. A bond breaker is used between floors. (page 227)
3. They provide necessary tensile strength and carry some of the compressive load. (page 229)
4. Any of the following: (page 241)
   - Deteriorated concrete
   - Spalling that exposes reinforcing rods
   - Cracks in concrete
5. Floors that are damaged on one side may give no indication as to the damage on the other side of the concrete. (page 244)

Word Fun

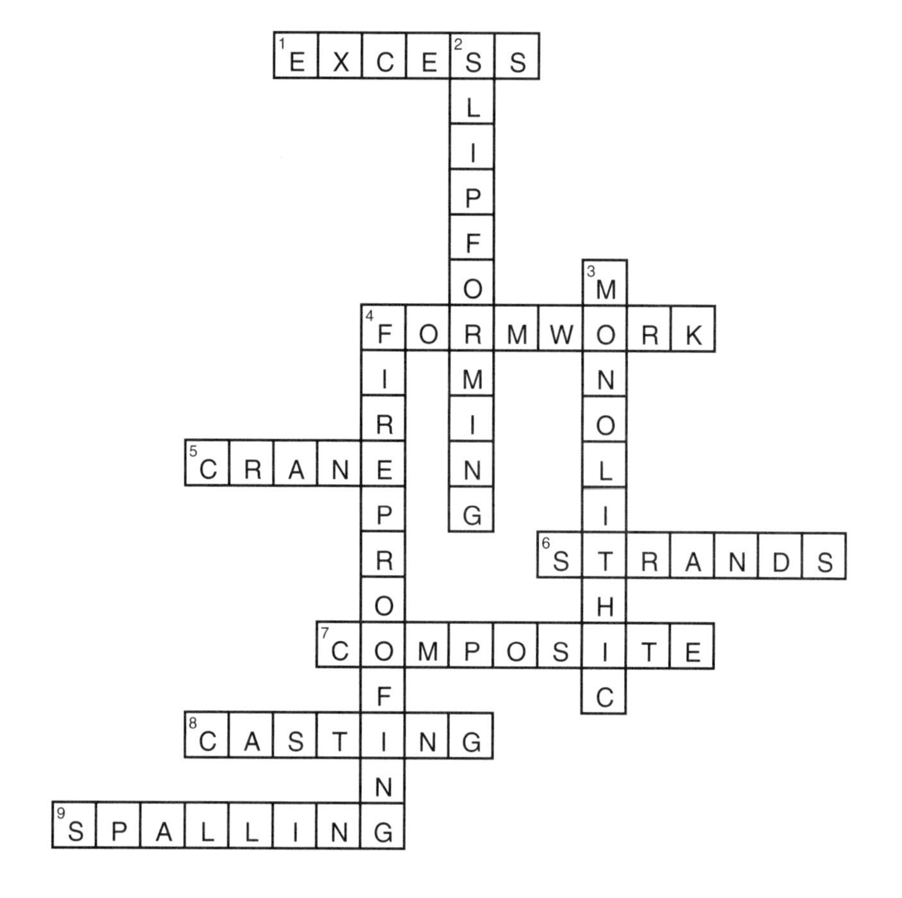

Across

1. Excess (page 236)

4. Formwork (page 233)

5. Crane (page 237)

6. Strands (page 231)

7. Composite (page 226)

8. Casting (page 226)

9. Spalling (page 248)

Down

2. Slipforming (page 228)

3. Monolithic (page 227)

4. Fireproofing (page 238)

Fire Alarms

1. During the construction process there are usually many workers. One of the first things to consider is how many people may be on site if something happens and whether there is accountability for them. Concrete can be very heavy, and a large number of strengthening rods may be encountered. Determine the tools that may be available to cut these reinforcing bars. Determine the types of forms that will be used and whether there will be any large pours. Sufficient personnel and tools must be brought to handle any collapse of the structure; fires that occur before the structure is fully cured may be very dangerous. (pages 225–237)

2. Any of the following: (pages 238–239)
   - Poor application of fireproofing
   - Holes that were not properly firestopped
   - Improperly supported ceilings
   - Equipment not properly supported from concrete
   - Spalling of concrete

Chapter 11: Specific Occupancy Details and Hazards

Matching

1. E (page 255)
2. C (page 265)
3. A (page 263)
4. B (page 252)
5. I (page 289)
6. J (page 269)
7. G (page 260)
8. H (page 277)
9. D (page 270)
10. F (page 270)

Multiple Choice

1. B (page 252)
2. D (page 252)
3. C (page 254)
4. B (page 253)
5. A (page 270)
6. D (page 271)
7. D (page 284)
8. C (page 255)
9. D (page 260)
10. B (page 260)

Fill-in-the-Blank

1. tenant (page 261)
2. nogging (page 259)
3. rows (page 258)
4. lapse (page 269)
5. proportional (page 270)
6. pressurized (page 273)
7. Cape Cod (page 280)
8. Victorian (page 281)
9. California (page 280)
10. ranch (page 281)

True/False

1. T (page 253)
2. T (page 252)
3. F (page 255)
4. T (page 255)
5. F (page 256)
6. T (page 257)
7. F (page 257)
8. F (page 262)
9. T (page 266)
10. F (page 267)

Short Answer

1. Any of the following: (page 267–268)
   - Common shafts to other elevators
   - Unreliable power
   - Detailed knowledge required to get people out of them
   - May have blind floors
2. Ignore the lock and go through the wallboard next to the door (page 269)
3. They prevent people from falling down a shaft by opening the doors in the wrong location. (page 269)
4. The upper floors are usually windier and the direction may change frequently. (page 270)
5. 1900 Hotel and Motel Fire Safety Act (page 277)

Word Fun

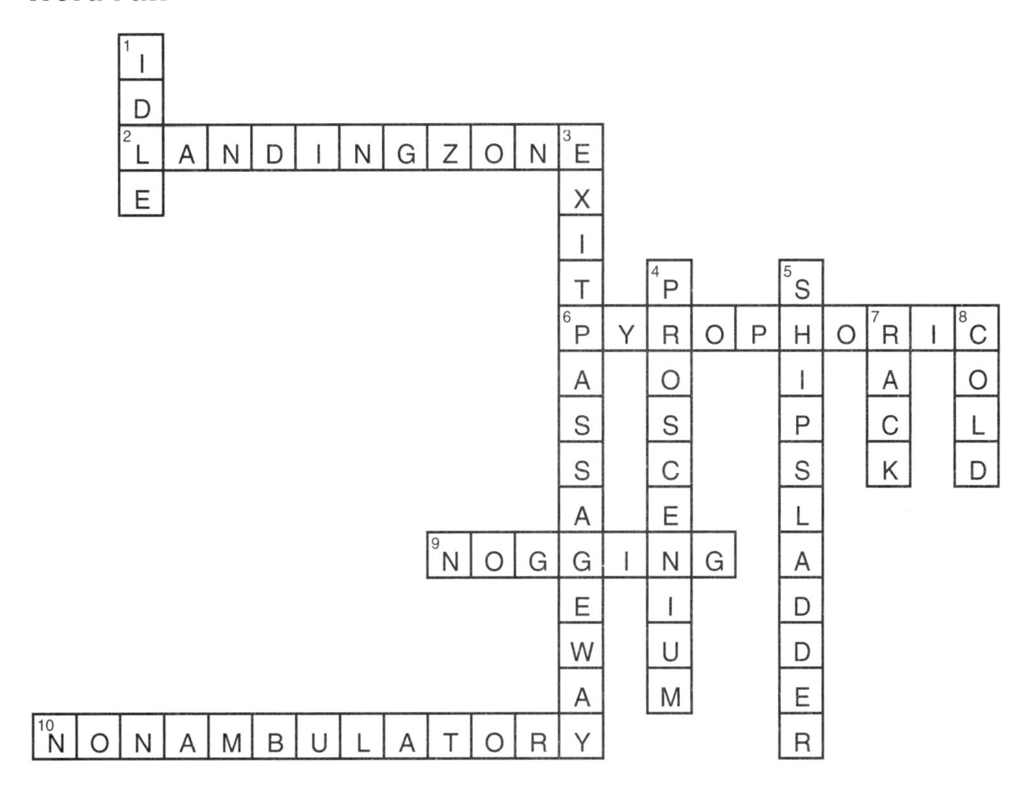

Across

2. Landing zone (page 269)
6. Pyrophoric (page 263)
9. Nogging (page 259)
10. Nonambulatory (page 276)

Down

1. Idle (page 287)
3. Exit passageway (page 262)
4. Proscenium (page 286)
5. Ship's ladder (page 262)
7. Rack (page 286)
8. Cold (page 271)

Fire Alarms

1. All of the following: (page 277)
   - Remove all people in immediate danger to safety.
   - Active the manual pull station and call 911.
   - Close doors to confine the spread of fire.
   - Extinguish the fire, if possible.

2. Any of the following: (pages 286–292)
   - Potential for collapse
   - Large fires
   - Hazardous materials
   - Racks that may fall
   - The possibility of getting disoriented

Chapter 12: Collapse

Matching

1. C (page 300)
2. H (page 300)
3. D (page 300)
4. A (page 300)
5. J (page 300)
6. B (page 301)
7. F (page 301)
8. G (page 301)
9. E (page 301)
10. I (page 301)

Multiple Choice

1. B (page 300)
2. D (page 300)
3. A (page 302)
4. B (page 302)
5. C (page 303)

Labeling

1. Pancake collapse (page 302)
2. Lean-to-floor collapse (page 302)
3. V-shaped floor collapse (page 302)
4. Curtain fall wall collapse (page 302)
5. Lean-over collapse (page 302)
6. Frame floor collapse (page 302)
7. 90-degree wall collapse (page 302)
8. Cantilever floor collapse (page 302)
9. Inward outward collapse (page 302)